# Lecture Notes in Computer Science 12334

More information about this series at http://www.springer.com/series/7412

Leszek J. Chmielewski · Ryszard Kozera ·
Arkadiusz Orłowski (Eds.)

# Computer Vision
# and Graphics

International Conference, ICCVG 2020
Warsaw, Poland, September 14–16, 2020
Proceedings

 Springer

*Editors*
Leszek J. Chmielewski ⓘ
Institute of Information Technology
Warsaw University of Life
Sciences - SGGW
Warsaw, Poland

Ryszard Kozera ⓘ
Institute of Information Technology
Warsaw University of Life
Sciences - SGGW
Warsaw, Poland

Arkadiusz Orłowski ⓘ
Institute of Information Technology
Warsaw University of Life
Sciences - SGGW
Warsaw, Poland

ISSN 0302-9743         ISSN 1611-3349   (electronic)
Lecture Notes in Computer Science
ISBN 978-3-030-59005-5         ISBN 978-3-030-59006-2   (eBook)
https://doi.org/10.1007/978-3-030-59006-2

LNCS Sublibrary: SL6 – Image Processing, Computer Vision, Pattern Recognition, and Graphics

This Springer imprint is published by the registered company Springer Nature Switzerland AG
The registered company address is: Gewerbestrasse 11, 6330 Cham, Switzerland

# Preface

The International Conference on Computer Vision and Graphics (ICCVG), organized since 2002, is the continuation of the International Conferences on Computer Graphics and Image Processing (GKPO), held in Poland every second year from 1990 to 2000. The main objective of ICCVG is to provide an environment for the exchange of ideas between researchers in the closely related domains of computer vision and computer graphics.

ICCVG 2020 brought together 50 authors. The proceedings contain 20 papers, each accepted on the grounds of merit and relevance confirmed by three independent reviewers. The conference was organized as an Internet event during the COVID-19 pandemic, which greatly reduced the attendance. Still, we maintained high-quality standards in the reviewing process.

ICCVG 2020 was organized by the Association for Image Processing, Poland (Towarzystwo Przetwarzania Obrazów – TPO) and the Institute of Information Technology at the Warsaw University of Life Sciences (SGGW), together with the Faculty of Information Science at the West Pomeranian University of Technology (WI ZUT) and the Polish-Japanese Academy of Information Technology (PJATK) as the supporting organizers.

The Association for Image Processing integrates the Polish community working on the theory and applications of computer vision and graphics. It was formed between 1989 and 1991.

The Institute of Information Technology was formed in 2019 as the result of structural changes in Polish science. It emerged from the structure of the Faculty of Applied Informatics and Mathematics (WZIM), established in 2008 at SGGW. Its location at the leading life sciences university in Poland is the source of opportunities for valuable research at the border of applied information sciences, forestry, furniture and wood industry, veterinary medicine, agribusiness, and the broadly understood domains of biology and economy.

We would like to thank all the members of the Scientific Committee, as well as the additional reviewers, for their help in ensuring the high quality of the papers. We would also like to thank Grażyna Domańska-Żurek for her excellent work on technically editing the proceedings, and Dariusz Frejlichowski, Krzysztof Lipka, Tomasz Minkowski, and Grzegorz Wieczorek for their engagement in the conference organization and administration.

September 2020

Leszek J. Chmielewski
Ryszard Kozera
Arkadiusz Orłowski

# Organization

Association for Image Processing (TPO), Poland
Faculty of Applied Informatics and Mathematics, Warsaw University of Life Sciences
  (WZIM SGGW), Poland
Polish-Japanese Academy of Information Technology (PJATK), Poland
Faculty of Computer Science and Information Technology, West Pomeranian
  University of Technology (WI ZUT), Poland
Springer, Lecture Notes in Computer Science (LNCS), Germany

## Conference General Chairs

Leszek J. Chmielewski, Poland
Ryszard Kozera, Poland
Arkadiusz Orłowski, Poland
Konrad Wojciechowski, Poland

## Scientific Committee

Ivan Bajla, Slovakia
Wojciech Maleika, Poland
Gunilla Borgefors, Sweden
Witold Malina, Poland
Nadia Brancati, Italy
Krzysztof Małecki, Poland
M. Emre Celebi, USA
Radosław Mantiuk, Poland
Leszek Chmielewski, Poland
Tomasz Marciniak, Poland
Dmitry Chetverikov, Hungary
Andrzej Materka, Poland
Piotr Czapiewski, Poland
Nikolaos Mavridis, UAE
László Czúni, Hungary
Przemysław Mazurek, Poland
Silvana Dellepiane, Italy
Tomasz Mąka, Poland
Marek Domański, Poland
Wojciech Mokrzycki, Poland
Mariusz Flasiński, Poland
Mariusz Nieniewski, Poland

Paweł Forczmański, Poland
Sławomir Nikiel, Poland
Dariusz Frejlichowski, Poland
Lyle Noakes, Australia
Maria Frucci, Italy
Antoni Nowakowski, Poland
André Gagalowicz, France
Adam Nowosielski, Poland
Duncan Gillies, UK
Krzysztof Okarma, France
Samuel Morillas Gómez, Spain
Maciej Orkisz, France
Ewa Grabska, Poland
Arkadiusz Orłowski, Poland
Diego Gragnaniello, Italy
Henryk Palus, Poland
Marcin Iwanowski, Poland
Wiesław Pamuła, Poland
Adam Jóźwik, Poland
Volodymyr Ponomaryov, Mexico
Heikki Kälviäinen, Finland
Piotr Porwik, Poland

# Contents

# Facial Age Estimation Using Compact Facial Features

Joseph Damilola Akinyemi$^{(\boxtimes)}$ and Olufade Falade Williams Onifade

Department of Computer Science, University of Ibadan, Ibadan, Nigeria
akinyemijd@gmail.com, olufadeo@gmail.com

**Abstract.** Facial age estimation studies have shown that the type of features used for face representation significantly impact the accuracy of age estimates. This work proposes a novel method of representing the face with compact facial features derived from extracted raw image pixels and Local Binary Patterns (LBP) for age estimation. The compact facial features are realized by exploiting the statistical properties of extracted facial features while aggregating over a whole feature set. The resulting compact feature set is of reduced dimensionality compared to the non-compact features. It also proves to retain relevant facial features as it achieves better age prediction accuracy than the non-compact features. An age group ranking model is also proposed which further reduces facial features dimensionality while improving age estimation accuracy. Experiments on the publicly-available FG-NET and Lifespan datasets gave a Mean Absolute Error (MAE) of 1.76 years and 3.29 years respectively, which are the lowest MAE so far reported on both datasets, to the best of our knowledge.

**Keywords:** Age estimation · Age-group ranking · Features compaction · Features extraction · Image analysis

## 1 Introduction

Research in Image Processing and Analysis is impacted by the kind of features employed. The statistical properties of the extracted features; such as its dimesionality and its discriminative power are some of the major factors that determine the success or failure of features at a particular image analysis task. Li et al. in [19] noted that automatic face recognition algorithms have improved over the years due to improved feature representation and classification models. Facial age estimation, being an image analysis task, has gained increasing research attention due to its applicability in homeland security and surveillance, age-specific human computer interaction and automated regulation of Internet contents [4,11].

Facial age estimation is impacted by the type of features employed for face representation. Some of the features descriptors employed for face representation in existing facial age estimation literature include Active Appearance Model

© Springer Nature Switzerland AG 2020
L. J. Chmielewski et al. (Eds.): ICCVG 2020, LNCS 12334, pp. 1–12, 2020.
https://doi.org/10.1007/978-3-030-59006-2_1

(AAM) [22], Local Binary Patterns (LBP) [6,9], Biologically-Inspired Features (BIF) [12] and Gabor Filters [5]. In [9], Eidinger et al. employed Local Binary Patterns (LBP) and Four-patch LBP in a drop-out SVM learning framework to estimate the ages and gender of faces in their Adience dataset. In [14], Biologically Inspired Features (BIF) were extracted from the face and its surrounding context in order to estimate age, gender and race from face images obtained in unconstrained environments. A multi-task warped Gaussian process was proposed as an extension of the Gaussian process in [29] for learning personalized ages and AAM features were used to learn common and person-specific features in an age regression function. In [10], BIF was extended by employing Active Shape Models (ASM) and the extended BIF (EBIF) used as facial features for age learning. BIF was also extracted from facial components and holistic face in [15] for age learning in a hierarchical age learning framework. A Grouping-Estimation-Fusion model for age estimation was proposed in [21] using Gradient Orientation Pyramid (GOP) for face representation. In [1], ethnic-specific age ranked features obtained from raw pixels were used in an ensemble of regressors to learn ages. Image pixels were also used in [20] in an Ordinal deep feature learning model for age estimation.

However, it has been observed that most works have employed features with high dimensionality e.g. [5,12,14] others employed a combination of two or more features yielding yet higher feature dimensions and increasing computational complexity e.g. [10,21] and others which employed deep learning models with high computational demands e.g. [8,20,25,27].

In a bid to reducing features dimensionality without losing pertinent discriminatory features, we propose a model of compacted features for facial age estimation. The proposed model aggregates extracted features over ranges of feature values rather than individual feature values, thereby creating a compact but discriminatory feature vector. The proposed compact features model does not only reduce the dimensionality of features but also captures statistical properties of the features which enhances the discriminatory properties of the derived compact features. The compacted features are employed in an age group ranking model which is employed for age estimation.

The rest of this paper is organized as follows; Sect. 2 describes the proposed model, Sect. 3 describes the experiments and Sect. 4 discusses the results obtained from the experiments while Sect. 5 concludes the paper.

## 2   The Proposed Model

In this work, a method of compacting extracted facial image features for the purpose of facial image processing – specifically, facial age estimation –is proposed. An overview of the age estimation model employed in this work is shown in Fig. 1.

We consider the age estimation problem as the task of labelling a face image with an age label. Thus, suppose we have a set $A = \{ a_1, \ldots a_p \}$ of face images and a set $B = \{ b_1, \ldots b_q \}$ of age labels, the goal of age estimation is to learn a function which appropriately maps $A$ to $B$. In order to do this, features must be obtained from face images which are believed to be responsible for the relationship between faces and their ages. Facial features are derived from a set of face regions defined as in (1), where $\ell$ is an arbitrary number of face regions.

$$R = \{r_k | k = 1...\ell\} \tag{1}$$

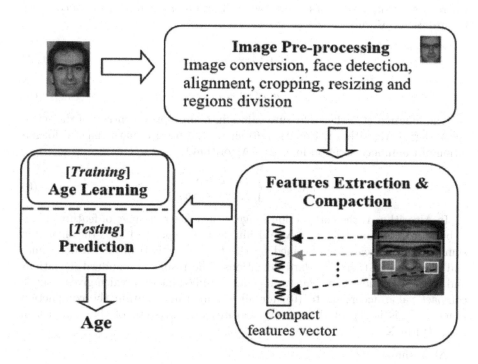

**Fig. 1.** Overview of the proposed facial age estimation approach.

## 2.1   Features Compaction Model

The intuition of compact features is this; since image pixels (picture elements) are low-level features that can be readily obtained from an image at relatively low computational cost, employing these features in a simple statistical model will not only reduce computational complexity, but also enhance the reproducibility of results. Therefore, the proposed features compaction algorithm seeks to represent image features in very low dimension without losing pertinent feature details.

To define the method of obtaining compact features, we start by defining a vector $\mathbf{X}$ of features $x_i$ and define a function $g(x_i, \tau)$ which obtains compacts feature $v_j \in \mathbf{V}$ (compact features vector) based on a specified compaction range,

$\tau$. The range ($\tau$) specifies the size of the subset of $\mathbf{X}$ to be compacted into each $v_j \in \mathbf{V}$.

$$\mathbf{X} = \{x_i | i = 1, ..., n\} \qquad (2)$$

$$\mathbf{V} = \{v_j | j = 1, ..., m\} \qquad (3)$$

where $n$ is the dimension of feature vector $\mathbf{X}$ and $m$ is the dimension of compact feature vector $\mathbf{V}$ and

$$m = \left\lceil \frac{n}{\tau} \right\rceil, 1 < \tau < n \qquad (4)$$

$$g(x_i, \tau) = v_j \qquad (5)$$

The function $g()$ collects feature values into appropriate entries of the set $\mathbf{V}$ according to Algorithm 1. Equation (6) shows the precise mathematical formulation of the function $c()$ on line 12 of Algorithm 1.

$$c(a, b) = \begin{cases} ab, \text{if } b < 1 \\ \frac{a}{b}, \text{otherwise} \end{cases} \qquad (6)$$

In Algorithm 1, the compaction range, $\tau$ specifies the range of feature values to be aggregated into each $v_j \in \mathbf{V}$. Within the range specified by $\tau$, the algorithm sums the frequencies (as obtained by the *freq()* function) of all feature values and *sd()* obtains their standard deviations. The resulting standard deviations and sums of frequencies within a range are supplied into $c()$ which gives a single compact value according to (6), for all feature values within the compaction range $\tau$. This is repeated for every feature value in steps of the compaction range ($\tau$) in $\mathbf{X}$.

Algorithm 1.
**Input:** Feature vector ($\mathbf{X}$), range ($\tau$), Feature dimension ($n$)
**Output:** Compact feature vector ($\mathbf{V}$)
1. $m = \left\lceil \frac{n}{\tau} \right\rceil$
2. for $i = 1$ to $m$
3.    $\mu[i] = 0$
4.    $k = ((i - 1) * \tau) + 1$
5.    for $q = k$ to $(k + \tau - 1)$
6.       if $(q < n)$
7.          $u[q] =$ freq($\mathbf{X}[q], \mathbf{X}$) //Get frequency of this feature value in $\mathbf{X}$
8.          $\mu[i] = \mu[i] + u[q]$
9.       end if
10.    end for
11.    $\sigma[i] =$ sd($u[k] ... u[k + \tau - 1]$ ) //Get std within $\tau$ range of ftrs
12.    $\mathbf{V}[i] = c(\mu, \sigma)$
13. end for
14. return $\mathbf{V}$

The resulting compacted features set is therefore of size $m$ which reduces with increasing size of $\tau$. As simple as this seems, its dimensionality reduction effect is very significant. For instance, in our experiments, facial images were divided into local regions from which feature sets were obtained. Each feature vector from each region is expected to contain the same range of feature values i.e. $x_1, ..., x_n$ where $n$ is the number of unique feature values (e.g. $n = 256$ for an 8-bit grayscale image) in $\mathbf{X}$ as seen in (2). This means $r \cdot n$ features will be obtained from the whole face (where r is the number of regions); these amount of features is huge especially when many regions are used. In this case, the compacted features model reduces the amount of features to $r \cdot m$ and $m$ is considerably less than $n$, according to (4). Apart from the dimensionality reduction effect of the proposed compact features model, it also ensures that statistical properties of the feature set are preserved by factoring in the standard deviation of features within each compaction range.

## 2.2   Age Learning

Age learning is here formulated as a supervised learning task (precisely, regression) which maps face features to age labels (real/ground truth ages). Therefore, the goal is to estimate a regression function which learns the relationship between faces and their ground truth ages. However, prior to age learning, face images are ranked by age group in order to embellish each face embedding with its correlation across various age groups. We adopted the age-group ranking model of [1] with the modifications that the entire training set, rather than a subset of it, is used as the reference image set. By using the entire training set as the reference image set for age-group ranking, the reference set is, therefore not necessarily individualized. However, to ensure that the age group ranks are not biased by the features of the image being ranked, each image (e.g. image $a_1$) in the training set is excluded from the reference set when that image (i.e. image $a_1$) is being ranked. The age group ranking function is defined as in (7).

$$\varphi_j \leftarrow h(a_i, \hat{a}_j, \iota) \tag{7}$$

where $a_i$ is the image to be ranked, $\hat{a}_j$ is a subset of $A$ (the set of images) containing images in the age group $j$ and $\iota$ represents the index of $a_i$ in $\hat{a}_j$, which indicates where to find the image to be ranked in the age group $j$ so that it can be exempted from being used as part of the reference image set. If the image is not present in the age group, a flag (e.g. $-1$) is set to indicate that and the same flag can be used for test images since they are not expected to be part of the training set. The age group ranking function in (7) is employed on all age groups for each image ai so that each image is completely ranked by age group according to (8)

$$\Phi_i \leftarrow h(a_i, \hat{a}_1, \iota) \oplus \ldots \oplus h(a_i, \hat{a}_s, \iota) \tag{8}$$

where $\Phi_i$ is the vector of age group ranks $(\varphi_i)$ of $a_i$ at each age group $j$, represents the operator which concatenates values into a vector and $s$ represents the number of age groups.

The obtained vector of age group ranks is then attached to the image embedding and used as the input into a regressor to learn the correlation between the supplied features and the ages.

## 3 Experiments and Results

The proposed model was experimented with Matlab R2016a. Our experiments were run on two publicly available datasets; FG-NET [7] and Lifespan [23] using Leave-One-Person-Out (LOPO) cross validation protocol on the former and 5-fold cross validation protocol on the latter as used in [13,21]. FG-NET contains 1002 face images of 82 Caucasian individuals at different ages spanning 0–69 years, while Lifespan contains 1046 images of 576 individuals of varying ethnicities within the age range 18–93 years. FG-NET and Lifespan datasets were subdivided into 11 and 12 age groups respectively (i.e. $s = 11$ and $s = 12$), which means 11 and 12 age group rank values for each image in FG-NET and Lifespan datasets respectively. Our choice of these datasets is due to their long-standing acceptance as benchmark datasets for age estimation, the individualized ageing patterns that can be learnt from a dataset like FG-NET and the fact that they are freely accessible to the public for research purposes.

The compact features model was experimented on raw pixel features and LBP features [24] (with 8 neighbours and 1 pixel radius) which were obtained from preprocessed face images. Preprocessing involved image conversion to grayscale, face alignment and detection using the method in [2] and aligned faces were cropped and rescaled to $120 \times 100$ pixels. Motivated by [16,17], each face image was divided into 10 local regions defined around the forehead, the outer eye corners, the inner eye corners, the area under the eyes, the area between the two eyes, the nose bridge, the nose lines, the cheek area, the cheek bone areas and the periocular face region. Thus, $\ell = 10$ as in (1) Compact features were obtained for 12 different compaction ranges $(\tau)$ (i.e. 5 to 60) as shown in Table 1. Support Vector Regression (SVR) with radial basis function kernel was employed for age learning, with box-conxtraint of 1 (for regularization). LBP and raw pixel features were chosen because of their simplicity and for ease of reproducibility of this work. Previous methods of deriving discriminatory features from LBP and image pixel are usually by simple histograms or selecting regions (or pixels) of maximum/average intensity. The derived compact features developed in this work additionally computes statistical relationships between feature patterns so that it increases the discriminatory properties of the features while reducing its dimensionality.

Age estimation performance is commonly evaluated by the Mean Absolute Error (MAE) and/or the Cumulative Score (CS). MAE is the average of absolute differences between the actual ages and the predicted ages while CS is a measure of the percentage of the images which are correctly estimated at given error levels.

Thus, the lower the MAE, the better the age estimation accuracy; the higher the CS, the better the age estimation accuracy.

## 4  Discussion

In this section, we discuss our findings from the experiments and explain the details in the results tables and figures. Note that the uniform LBP features used throughout our experiments is as described in [24]. In all cases, features were extracted and compacted from each face region and then features from all regions were fused into a vector. LBP features were compacted in two different manners. In the first method, LBP values were first computed over the whole face, then the computed LBP features are compacted from each region while the second method computes LBP values within each region (not from the whole face) and compacts the resulting LBP features from each region. The first method is referred to as whole face (WF), while the second method is called face regions (FR). In all tables and figures, ULBP, NC and RP stand for Uniform LBP features, Non-Compact features and Raw Pixel features respectively. Also, all reported results were obtained via cross validated training and testing using LOPO on FG-NET dataset and 5-fold on Lifespan dataset.

**Table 1.** MAE of compact versus non-compact features

| Value of $\tau$ | # of ftrs. | MAE (years) | | | | | |
|---|---|---|---|---|---|---|---|
| | | FG-NET | | | Lifespan | | |
| | | RP | LBP | | RP | LBP | |
| | | | WF | FR | | WF | FR |
| 5 | 520 | 8.42 | 7.66 | 7.82 | 13.99 | 11.32 | 11.25 |
| 10 | 260 | 8.11 | 6.84 | 6.40 | 13.05 | 10.57 | 9.87 |
| 15 | 180 | **7.98** | 5.74 | 5.40 | **12.71** | 9.26 | 9.34 |
| 20 | 130 | 8.10 | 7.82 | 7.59 | 12.80 | 9.39 | 9.51 |
| 25 | 110 | 8.15 | 6.75 | 6.57 | 12.55 | **8.95** | **8.89** |
| 30 | 90 | 8.09 | 6.27 | 5.63 | 12.72 | 9.68 | 8.99 |
| 35 | 80 | 8.08 | 7.75 | 7.74 | 12.72 | 9.81 | 9.88 |
| 40 | 70 | 8.41 | 5.78 | 6.09 | 13.33 | 9.25 | 9.45 |
| 45 | 60 | 8.13 | 7.02 | 7.09 | 13.14 | 10.51 | 10.23 |
| 50 | 60 | 8.28 | 6.48 | 5.90 | 13.69 | 9.06 | 9.00 |
| 55 | 50 | 8.19 | 6.21 | 6.57 | 13.48 | 9.70 | 9.53 |
| 60 | 50 | 8.25 | **5.59** | **5.08** | 13.28 | 10.67 | 9.75 |
| ULBP | 590 | – | 7.35 | 7.32 | – | 10.41 | 10.33 |
| NC | 2560 | 8.77 | 8.10 | 8.08 | 14.79 | 11.87 | 11.73 |

As seen in Table 1, on both datasets, compact raw pixel features generally performed better than non-compact raw pixel and uniform LBP features. On FG-NET, the lowest MAE values are recorded at compaction ranges 15 for raw pixel and 60 for LBP while compaction range 25 gives the lowest error on Lifespan dataset. Our inference from this is that different compaction ranges offer more advantage for better performance than non-compact features. More so, considering the huge dimensionality of non-compact features (2560 features) which is almost 5 times the size of the compact features with the highest dimension of just 520 features (at compaction range of 5), the performance gain offered by compact features is quite significant.

Table 2 records age estimation performance obtained by using only the obtained age group ranks as in (8). This means that features with dimensionalities 11 and 12 are used for age estimation on FG-NET and Lifespan datasets respectively. Table 2 shows that nearly the same trend of performance is retained on both datasets with LBP features generally performing better than raw pixel features. At this point, non-compact raw pixel features outperform all compact raw pixel features on both datasets. This shows that with non-compact raw pixel features, images were better age-ranked than compact raw pixel features. This also shows how age group ranking could improve age learning and consequently age estimation accuracy. However, considering the difference between the dimensionality of compact and non-compact features (2560 features), it may be argued that the best performing compact raw pixel features with MAE of 3.91 years ($\tau = 10$) performs at par with non-compact raw pixels. More so, the significant improvement of compact over non-compact features is confirmed by the performance of LBP features (both WF and FR) as shown in Table 2, with non-compact LBP features being outperformed by compact LBP features at about 6 different compaction ranges (at $\tau = 5, 10, 15, 30, 40$ and 60) on FG-NET dataset and about 3 different compaction ranges (at $\tau = 5, 10$ and 15) on Lifespan dataset. Therefore, for LBP features, compaction seems to be a better method of collecting/aggregating features than the usual histogram or uniform patterns.

Figure 2 shows the CS of our proposed method in comparison with those reported in literature on FG-NET dataset. C15-AGR stands for compact features ($\tau = 15$) with age group rankings. The CS along with the MAE usually allows for performance comparison between different models and datasets because the CS shows how well an age estimation algorithm performs at different error levels. The plot shows the rank of compact features is at par with the state-of-the-art algorithms with the compact LBP features starting out at over 20% at zero-error level and competing well until it outperforms other algorithms from error-level 3 upwards. In the same vein, Table 3 shows the various MAE values obtained from previous models as well as the proposed model on FG-NET and Lifespan datasets and our proposed age group ranks of compact features outperformed the state-of-the-art algorithms in facial age estimation on both datasets. It is noteworthy that most of the state-of-the-art algorithms shown in Fig. 2 and Table 3 employed features with dimensionality in their hundreds (e.g. DeepRank

**Table 2.** MAE of age-group-ranked compact versus non-compact features

| Value of $\tau$ | MAE (years) | | | | | |
|---|---|---|---|---|---|---|
| | FG-NET | | | Lifespan | | |
| | RP | LBP | | RP | LBP | |
| | | WF | FR | | WF | FR |
| 5 | 5.03 | 2.27 | 2.26 | 9.84 | **3.69** | 3.41 |
| 10 | 3.91 | **1.88** | 1.79 | 9.23 | 4.03 | **3.29** |
| 15 | 4.42 | 2.02 | **1.76** | 9.24 | 4.68 | 3.99 |
| 20 | 4.67 | 3.80 | 3.24 | 10.33 | 7.05 | 7.64 |
| 25 | 4.98 | 3.27 | 3.01 | 11.87 | 5.67 | 5.44 |
| 30 | 5.36 | 2.67 | 2.36 | 11.46 | 6.65 | 5.79 |
| 35 | 5.65 | 4.81 | 4.65 | 11.64 | 8.44 | 8.77 |
| 40 | 6.02 | 2.96 | 3.19 | 12.99 | 7.06 | 8.26 |
| 45 | 6.00 | 4.79 | 4.63 | 14.39 | 9.22 | 8.46 |
| 50 | 6.95 | 4.05 | 3.53 | 13.23 | 8.43 | 9.48 |
| 55 | 6.53 | 4.01 | 4.27 | 13.35 | 9.21 | 9.35 |
| 60 | 7.24 | 2.93 | 2.66 | 13.67 | 8.56 | 5.95 |
| ULBP | – | 4.37 | 4.39 | – | 6.60 | 6.26 |
| NC | **2.95** | 3.19 | 3.12 | **9.00** | 4.95 | 4.53 |

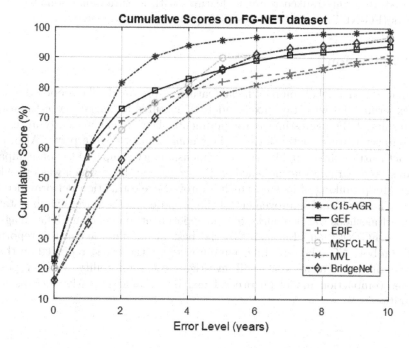

**Fig. 2.** Cumulative Scores (CS) on FG-NET using LOPO.

**Table 3.** Comparison of MAE on FG-NET (LOPO) and Lifespan (5-fold)

| Method | MAE (years) | |
|---|---|---|
| | FG-NET | Lifespan |
| ODFL [20] | 3.89 | – |
| EBIF [10] | 3.17 | – |
| GEF [21] | 2.81 | – |
| MSFCL-KL [26] | 2.71 | – |
| MVL [25] | 2.68 | – |
| BridgeNet [18] | 2.56 | – |
| BPM-2 [13] | – | 6.19 |
| W-RS [28] | – | 5.99 |
| Joint-Learn [3] | – | 5.26 |
| DeepRank [27] | – | 4.31 |
| Transfer Learning [8] | – | 3.79 |
| C15-LBP-FR | **1.76** | **3.99** |
| C10-LBP-FR | **1.79** | **3.29** |

[27] used 500 features) and some even combined two to three features (e.g. GEF [21] used LBP, BIF and HOG). Therefore, the performance improvement of our proposed age-group-ranked compact features with feature dimensionality of 11 features (FG-NET) and 12 features (Lifespan) is very significant.

## 5   Conclusion

In this work, a model of compact facial features is proposed. The proposed model was experimented with raw pixel and LBP features for facial age estimation. The results of the age estimation experiments showed that in nearly all cases, the compact features outperformed their non-compact counterparts. Also, the compact features are of highly reduced dimensionality compared to non-compact features. Age learning was modelled as a regression problem which was preceded by age group ranking of images which improved age estimation performance by ≈40%. The significant improvement of the proposed method over existing state-of-the-art methods was demonstrated via experiments on benchmark facial aging datasets, FG-NET and Lifespan. To the best of our knowledge, the reported MAEs on both FG-NET and Lifespan datasets are the lowest reported on these datasets till date. Future works will investigate the applicability of the proposed features compaction model on other facial image analysis tasks such as face recognition.

# References

1. Akinyemi, J.D., Onifade, O.F.W.: An ethnic-specific age group ranking approach to facial age estimation using raw pixel features. In: 2016 IEEE Symposium on Technologies for Homeland Security (HST), pp. 1–6. IEEE, Waltham (May 2016). https://doi.org/10.1109/THS.2016.7819737

2. Akinyemi, J.D., Onifade, O.F.W.: A computational face alignment method for improved facial age estimation. In: 2019 15th International Conference on Electronics, Computer and Computation, ICECCO 2019. No. ICECCO, IEEE, Abuja (2019). https://doi.org/10.1109/ICECCO48375.2019.9043246

3. Alnajar, F., Alvarez, J.: Expression-invariant age estimation. In: Valstar, M., French, A., Pridmore, T. (eds.) British Machine Vision Conference (BMVC), pp. 1–11. BMVA Press (2014). https://doi.org/10.5244/C.28.14

4. Angulu, R., Tapamo, J.R., Adewumi, A.O.: Age estimation via face images: a survey. EURASIP J. Image Video Process. **2018**(42), 1–35 (2018). https://doi.org/10.1186/S13640-018-0278-6

5. Chen, C., Yang, W., Wang, Y., Shan, S., Ricanek, K.: Learning gabor features for facial age estimation. In: Sun, Z., Lai, J., Chen, X., Tan, T. (eds.) CCBR 2011. LNCS, vol. 7098, pp. 204–213. Springer, Heidelberg (2011). https://doi.org/10.1007/978-3-642-25449-9_26

6. Choi, S.E., Lee, Y.J., Lee, S.J., Park, K.R., Kim, J.: Age estimation using a hierarchical classifier based on global and local facial features. Pattern Recogn. **44**(6), 1262–1281 (2011). https://doi.org/10.1016/j.patcog.2010.12.005

7. Cootes, T.F., Rigoll, G., Granum, E., Crowley, J.L., Marcel, S., Lanitis, A.: FG-NET: Face and Gesture Recognition Working group (2002). http://www-prima.inrialpes.fr/FGnet/

8. Dornaika, F., Arganda-Carreras, I., Belver, C.: Age estimation in facial images through transfer learning. Mach. Vis. Appl. **30**(1), 177–187 (2019). https://doi.org/10.1007/s00138-018-0976-1

9. Eidinger, E., Enbar, R., Hassner, T.: Age and gender estimation of unfiltered faces. IEEE Trans. Inf. Forensics Secur. **9**(12), 2170–2179 (2014). https://doi.org/10.1109/TIFS.2014.2359646

10. El Dib, M.Y., Onsi, H.M.: Human age estimation framework using different facial parts. Egypt. Inform. J. **12**(1), 53–59 (2011). https://doi.org/10.1016/j.eij.2011.02.002

11. Fu, Y., Guo, G., Huang, T.S.: Age synthesis and estimation via faces: a survey. IEEE Trans. Pattern Anal. Mach. Intell. **32**(11), 1955–1976 (2010). https://doi.org/10.1109/TPAMI.2010.36

12. Guo, G., Mu, G., Fu, Y., Dyer, C.R., Huang, T.S.: A study on automatic age estimation using a large database. In: 2009 IEEE 12th International Conference on Computer Vision, vol. 12, pp. 1986–1991 (2009). https://doi.org/10.1109/ICCV.2009.5459438, http://ieeexplore.ieee.org/document/5459438/

13. Guo, G., Wang, X.: A study on human age estimation under facial expression changes. In: 2012 IEEE Conference on Computer Vision and Pattern Recognition, pp. 2547–2553 (2012). https://doi.org/10.1109/CVPR.2012.6247972

14. Han, H., Jain, A.K.: Age, gender and race estimation from unconstrained face images. Technical report, Michigan State University, East Lansing, Michigan (2014). http://www.cse.msu.edu/rgroups/biometrics/Publications/Face/HanJain_UnconstrainedAgeGenderRaceEstimation_MSUTechReport2014.pdf

15. Han, H., Otto, C., Jain, A.K.: Age estimation from face images: human vs. machine performance. In: 2013 International Conference on Biometrics (ICB), pp. 1–8 (2013). https://doi.org/10.1109/ICB.2013.6613022, http://ieeexplore.ieee.org/document/6613022/

16. Lanitis, A.: On the significance of different facial parts for automatic age estimation. In: International Conference on Digital Signal Processing, DSP, vol. 2, no. 14, pp. 1027–1030 (2002). https://doi.org/10.1109/ICDSP.2002.1028265

17. Lemperle, G., Holmes, R.E., Cohen, S.R., Lemperle, S.M.: A classification of facial wrinkles. Plast. Reconstr. Surg. **108**(6), 1735–1750 (2001). https://doi.org/10.1097/00006534-200111000-00048

18. Li, W., Lu, J., Feng, J., Xu, C., Zhou, J., Tian, Q.: BridgeNet: a continuity-aware probabilistic network for age estimation. In: IEEE Computer Society Conference on Computer Vision and Pattern Recognition, pp. 1145–1154 (2019). http://arxiv.org/abs/1904.03358

19. Li, Z., Gong, D., Li, X., Tao, D.: Aging face recognition: a hierarchical learning model based on local patterns selection. IEEE Trans. Image Process. **25**(5), 2146–2154 (2016). https://doi.org/10.1109/TIP.2016.2535284

20. Liu, H., Lu, J., Feng, J., Zhou, J.: Ordinal deep feature learning for facial age estimation. In: Proceedings - 12th IEEE International Conference on Automatic Face and Gesture Recognition, FG 2017–1st International Workshop on Adaptive Shot Learning for Gesture Understanding and Production, ASL4GUP 2017, Biometrics in the Wild, Bwild 2017, Heterogeneous, pp. 157–164 (2017). https://doi.org/10.1109/FG.2017.28, http://ieeexplore.ieee.org/document/7961736/

21. Liu, K.H., Yan, S., Kuo, C.C.J.: Age estimation via grouping and decision fusion. IEEE Trans. Inf. Forensics Secur. **10**(11), 2408–2423 (2015). https://doi.org/10.1109/TIFS.2015.2462732

22. Luu, K., Ricanek, K., Bui, T.D., Suen, C.Y.: Age estimation using active appearance models and support vector machine regression. In: 2009 IEEE 3rd International Conference on Biometrics: Theory, Applications, and Systems, pp. 1–5 (2009). https://doi.org/10.1109/BTAS.2009.5339053

23. Minear, M., Park, D.C.: A lifespan database of adult facial stimuli. Behav. Res. Methods. Instrum. Comput. **36**(4), 630–633 (2004). https://doi.org/10.3758/BF03206543

24. Ojala, T., Pietikäinen, M., Mäenpää, T.: Multiresolution gray-scale and rotation invariant texture classification with local binary patterns. IEEE Trans. Pattern Anal. Mach. Intell. **24**(7), 971–987 (2002). https://doi.org/10.1109/TPAMI.2002.1017623

25. Pan, H., Han, H., Shan, S., Chen, X.: Mean-variance loss for deep age estimation from a face. In: Proceedings of the IEEE Computer Society Conference on Computer Vision and Pattern Recognition, pp. 5285–5294 (2018). https://doi.org/10.1109/CVPR.2018.00554

26. Xia, M., Zhang, X., Liu, W., Weng, L., Xu, Y.: Multi-stage feature constraints learning for age estimation. IEEE Trans. Inf. Forensics Secur. **15**(c), 2417–2428 (2020). https://doi.org/10.1109/TIFS.2020.2969552

27. Yang, H.F., Lin, B.Y., Chang, K.Y., Chen, C.S.: Automatic age estimation from face images via deep ranking. In: British Machine Vision Conference (BMVC), no. 3, pp. 1–11 (2015). https://doi.org/10.5244/C.29.55

28. Zhang, C., Guo, G.: Age estimation with expression changes using multiple aging subspaces. In: IEEE International Conference on Biometrics: Theory, Applications and Systems (BTAS), pp. 1–6. Arlington (2013). https://doi.org/10.1109/BTAS.2013.6712720

29. Zhang, Y., Yeung, D.Y.: Multi-task warped Gaussian process for personalized age estimation. In: 2010 IEEE Conference on Computer Vision and Pattern Recognition (CVPR), pp. 2622–2629 (2010). https://doi.org/10.1109/CVPR.2010.5539975

# A Vision Based Hardware-Software Real-Time Control System for the Autonomous Landing of an UAV

Krzysztof Blachut⬛, Hubert Szolc⬛, Mateusz Wasala⬛, Tomasz Kryjak$^{(\boxtimes)}$⬛, and Marek Gorgon⬛

Embedded Vision Systems Group, Computer Vision Laboratory Department of Automatic Control and Robotics, AGH University of Science and Technology, Krakow, Poland
{kblachut,szolc,wasala,tomasz.kryjak,mago}@agh.edu.pl

**Abstract.** In this paper we present a vision based hardware-software control system enabling the autonomous landing of a multirotor unmanned aerial vehicle (UAV). It allows for the detection of a marked landing pad in real-time for a $1280 \times 720$ @ 60 fps video stream. In addition, a LiDAR sensor is used to measure the altitude above ground. A heterogeneous Zynq SoC device is used as the computing platform. The solution was tested on a number of sequences and the landing pad was detected with 96% accuracy. This research shows that a reprogrammable heterogeneous computing system is a good solution for UAVs because it enables real-time data stream processing with relatively low energy consumption.

**Keywords:** Drone landing · UAV · FPGA · Zynq SoC · Embedded vision

## 1 Introduction

Unmanned Aerial Vehicles (UAV), commonly known as drones, are becoming increasingly popular in commercial and civilian applications, where they often perform simple, repetitive tasks such as terrain patrolling, inspection or goods delivering. Especially popular among drones are the so-called multirotors (quadrocopters, hexacopters etc.), which have very good navigation capabilities (vertical take-off and landing, hovering, flight in narrow spaces). Unfortunately, their main disadvantage is a high energy consumption. With the currently used Li-Po batteries the flight time is relatively short (up to several dozen minutes). Therefore, the autonomous realisation of the mentioned tasks requires many take-offs and landings to replace or recharge batteries.

The start of a multirotor, assuming favourable weather conditions (mainly the lack of the strong wind) and keeping the distance from the other obstacles, is simple. Landing in a selected place, however, requires relatively precise

© Springer Nature Switzerland AG 2020
L. J. Chmielewski et al. (Eds.): ICCVG 2020, LNCS 12334, pp. 13–24, 2020.
https://doi.org/10.1007/978-3-030-59006-2_2

navigation. If a tolerance of up to several meters is allowed, the GPS (Global Positioning System) signal can be used. Nevertheless, it should be noted that in the presence of high buildings the GPS signal may disappear or be disturbed. What is more, even under favourable conditions, the accuracy of the determined position is limited to approximately 1–2 m. Performing a more precise landing requires an additional system. There are primarily two solutions to this problem – the first one is based on the computer vision and the second one uses a radio signal to guide the vehicle.

The main contribution of this paper is the proposal of a hardware-software vision system that enables the precise landing of the drone on a marked landing pad. As the computational platform, a heterogeneous Zynq SoC (System on Chip) device from Xilinx is used. It consists of a programmable logic (PL or FPGA – Field Programmable Gate Array) and a processor system (PS) based on a dual-core ARM processor. This platform allows to implement video stream processing with $1280 \times 720$ resolution in real time (i.e. 60 frames per second), with relatively low power consumption (several watts). The processor facilitates the communication with the drone's controller and is responsible for the final part of the vision algorithm. To the best of our knowledge, this is the first reported implementation of such a system in a heterogeneous device.

The remainder of this paper is organised as follows. In Sect. 2 the previous works on an autonomous drone landing are briefly presented. Then, the multirotor used in the experiments, the landing procedure and the landing pad detection algorithm are described. In Sect. 4 the details of the created hardware-software system are presented. Subsequently, in Sect. 5, the evaluation of the designed system is discussed. The paper concludes with a summary and the discussion about the further research directions.

## 2   Previous Works

The use of the visual feedback from the UAV's camera to perform a specific task autonomously has been addressed in a number of scientific papers. One type of them focuses on a combination of different types of autonomous vehicles to perform the different tasks. One such work is [3], in which the ground vehicle (UGV) is responsible for transporting and recharging the drone, while the UAV is responsible for the analysis of the environment and the navigation for the UGV. However, in the following review we focus mainly on one task – the landing on a static pad for the UAV without the use of any other vehicle.

The authors of the work [5] reviewed different autonomous landing algorithms of a multirotor on both static and moving landing pads. They compared all previously used landing pad tags, i.e. ArUco, ARTag, and AprilTag, tags based on circles and a "traditional" H-shaped marker. They also pointed out that landing in an unknown area, i.e. without a designated landing pad, is a challenging task. Then they analysed two types of drone controllers during the landing phase: PBVS (Position-Based Visual Servoing) and IBVS (Image-Based Visual Servoing). The first compares a drone's position and orientation with the expected values. The second one compares the location of the feature points between the pattern and subsequent image frames. In the summary, the authors pointed out three main challenges related to this type of systems: the development of a reliable vision algorithm with limited computing resources, the improvement of a state estimation and the appropriate modelling of the wind in the control algorithm.

In the work [8] the authors proposed an algorithm for landing of an unmanned aerial vehicle on a stationary landing pad. It was used when replacing or recharging the drone's battery was necessary during high-voltage line inspection. For the detection of the landing pad they used thresholding, median filtering, contour extraction, geometrical moments and an SVM classifier. In addition, the Extended Kalman Filter (EKF) was used to determine the position of the drone. It processed data from the inertial sensors (IMU – Inertial Measurement Unit), radar and the vision system. During tests, the drone approached the landing pad with a GPS sensor and then switched to the vision mode, in which the landing pad was detected with a camera. Out of 20 sequences, 14 ended with a position error of less than 10 cm, and the remaining ones below 20 cm. At the same time, in 12 tests the orientation error was below $10°$, while in the remaining ones it was below $20°$. As a computing platform, the authors used Raspberry Pi. They obtained a processing time of a single frame of 0.19 s, that is about 5 frames per second, which is sufficient for slow-moving drones. No information about the camera resolution was provided.

The authors of the work [10] presented an algorithm for controlling the autonomous landing of an unmanned aerial vehicle on a stationary T-shaped landing pad. The proposed vision algorithm was based on so-called image key points (feature points). The algorithm consisted of colour to greyscale transformation, thresholding, morphological operations, contour extraction (by using the Laplace operator) and matching the polygon to the detected points. Based on this, 8 angular points were determined, which were used to find the position of the drone corresponding to the landing pad. The algorithm worked in real-time as the processing of one frame with a resolution of $480 \times 320$ pixels took about 50 ms. The authors did not state on which platform the algorithm was implemented.

The approach of using Deep Learning methods for UAVs is getting increasingly popular in recent years. The survey [1] presents the most typical methods and applications, mainly feature extraction, planning and motion control. Although using Deep Learning for autonomous landing is not very common, one of the articles [6] presents the possibility of using reinforcement learning to generate the control necessary for this task. Real experiments were also carried out on a hexacopter. It was equipped with a custom computer based on a TI microcontroller, NVIDIA Jetson TX2 platform and an HD camera. The authors reported a single frame processing time of 200 ms (5 fps). In addition, they mentioned that this was one of the reasons for the observed oscillations. It should be noted that the size of the Deep Learning models is usually too big for the embedded platforms that can be placed on a small UAV. Their another downside is that they may not operate in real-time, which is critical for the autonomous landing task.

The article [7] presents a complex vision algorithm for landing pad detection. A marker in the form of the letter H placed inside a circle with an additional smaller circle in the centre was used. The algorithm operation was divided into three phases that were performed depending on the distance between the drone and the landing pad. In the first one, the outer circle, then the letter H, and finally the middle circle were detected. The authors devoted much attention to the reliability of the algorithm, presenting good results in the conditions of partial occlusion and shading of the marker. An Odroid XU4 computing platform with a Linux operating system was used and the OpenCV library was applied. The source of the video stream was a camera with a resolution of $752 \times 480$ pixels. Processing of 12 to 30 frames per second was obtained depending on the phase of the considered algorithm.

In the work [4] the authors proposed the RTV (Relative Total Variation) method to filter the unstructured texture to improve the marker (a triangle inside a circle) detection reliability. They also used median filtering, Canny edge detection and polygon fitting by the Ramer-Douglas-Peucker algorithm. In addition, they proposed a method of integrating data obtained from the GPS sensor and vision algorithm. There was no information about the used computing platform, camera resolution or the possibility of performing calculations in real-time.

Summarising the review, it is worth noting that in most works vision systems were able to process just a few frames per second. The authors of [6] pointed out that that this could be the reason the drone was oscillating during landing. To support our observations, the authors in [5] also perceive as a challenge the use of an energy-efficient platform to perform calculations in real-time. In this work, we claim that the use of a heterogeneous computing platform can address the both mentioned issues.

## 3   The Proposed Automatic Landing Algorithm

In this section we present the used hardware setup, the proposed automatic landing procedure and the landing pad detection algorithm.

### 3.1   Hardware Setup

In this research we used a custom multirotor (hexacopter) built from the following components:

- DJI F550 frame, DJI 2312/960KV engines controlled by 420 LITE controllers and equipped with reinforced propellers, with the designation 9050, i.e. with a propeller diameter equal to 9.0" (22.86 cm) and a 5.0" pitch (12.7 cm),
- four-cell Li-Po battery with a nominal voltage of 14.8 V (maximum 16.8 V) and with a capacity of 6750 mAh,
- FrSky Taranis X9D Plus radio equipment with FrSky X8D receiver,
- 3DR Pixhawk autoilot.

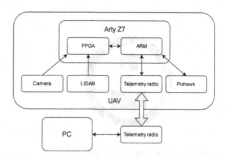

**Fig. 1.** The used hexacopter

**Fig. 2.** A simplified scheme of the proposed system

The drone was adapted to the considered project. We added a heterogeneous computing platform Arty Z7 with Zynq-7000 SoC device. We connected an Yi Action camera to it (as the source of a $1280 \times 720$ @ 60 fps video stream), a LiDAR for measuring altitude above the ground level (a LIDAR-Lite v3 device, SEN-14032) and a radio transceiver module for wireless communication with the ground station (3DR Mini V2 telemetry modules). Moreover, the Arty Z7 board was connected to the Pixhawk autopilot via UART (Universal Asynchronous Receiver-Transceiver). The drone is shown in Fig. 1, while the simplified functional diagram of the proposed system is presented in Fig. 2.

### 3.2 The Proposed Automatic Landing Procedure

In the initial situation, the drone flies at an altitude of about 1.5–2 m above the ground level and the landing pad is in the camera's field of view. Until then, the vehicle is piloted manually or in another automated mode (using GPS-based navigation). After fulfilling the conditions mentioned above, the system is switched into autonomous landing mode (currently implemented by a switch on the remote controller). However, if the described conditions are not met, the autonomous landing can not be performed. If the drone's altitude exceeds 2 m, the landing pad is not detected correctly because of the filtrations used to eliminate small objects.

In the first phase, a rough search for the landing pad is executed. After finding it, the drone autonomously changes its position so that the centre of the marker is around the centre of the image. In the second phase, the altitude is decreased to approx. 1 m and the drone's orientation relative to the landing pad is additionally determined. Based on this information, the vehicle is positioned accordingly.

In the last phase, the landing is carried out. The altitude above the landing pad is measured using the LiDAR sensor. The position of the pad is constantly determined and corrections, if required, are made. The drone is lowered and the engines are turned off after landing.

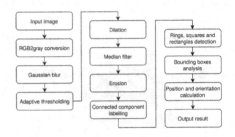

**Fig. 3.** The used landing marker

**Fig. 4.** Simplified dataflow of the vision algorithm

### 3.3   The Landing Spot Detection Algorithm

The key element of the proposed autonomous landing system is the landing pad detection algorithm. Based on the analysis of previous works and preliminary experiments, we decided to choose the landing pad marker as shown in Fig. 3. It consists of a large black circle, inside which we placed three other geometrical figures – a square, a rectangle and a small circle. The largest of the figures made it possible to roughly detect the landing pad and determine the approximate centre of the marker. The small circle was used in the case of a low altitude of the drone i.e. when the large circle was no longer fully visible in the image. It also allowed to improve the accuracy of the marker centre detection. The purpose of placing the square and the rectangle was to determine the orientation of the entire marker. Therefore, we are able to land the drone in a certain place with a given orientation. Here we should note, that the used shapes are relatively easy to detect in a vision system implemented in reconfigurable logic resources.

A simplified block diagram of the entire vision algorithm is depicted in Fig. 4. The application was prototyped in the C++ language with the OpenCV library version 4.1.0, which includes most basic image processing operations. We provide the source code of our application [9]. This allowed to compare different approaches, select parameters, etc. The obtained results could then be used during hardware implementation on the target platform.

Firstly, the recorded image is converted to greyscale. Then a standard Gaussian blur filtering is applied with a 5 × 5 kernel. In the next step adaptive thresholding is used. As a result of its use, the segmentation of the landing pad works well in different lighting conditions. The input image is divided into non-overlapping windows of 128 × 128 pixels. For each of them it is possible to determine both the minimum and the maximum brightness of the pixel. Then the thresholding is performed in which the threshold is calculated according to Eq. (1).

$$th = 0.5 \cdot (max - min) + min \tag{1}$$

where: *th* it the local threshold for the window, *max* is the maximum pixel brightness in the window and *min* is the minimum pixel brightness in the window.

The second stage of adaptive thresholding is the bilinear interpolation of the threshold's value. For each pixel, its four neighbours are determined (for pixels on the edges two neighbours, while in the corner just one) and the binarization threshold is calculated. As a result, the output binary image is smoother and the boundaries between the windows are not visible. This is presented in Fig. 5(d).

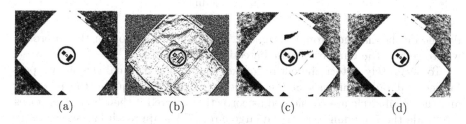

(a)                (b)                (c)                (d)

**Fig. 5.** A comparison of thresholding methods. (a) global thresholding, (b) local thresholding, (c) adaptive thresholding, (d) adaptive thresholding with interpolation.

It is worth noting that alternative binarization approaches were also considered, i.e. based on a global threshold – Fig. 5(a), "fully local" (the threshold for each pixel calculated on the basis of its local neighbourhood) – Fig. 5(b), and adaptive in non-overlapping windows but without an interpolation Fig. 5(c). Figure 5 presents a comparison of the described above approaches performed on a sample image. It can be concluded that the adaptive interpolation method is the best choice for the considered system. It provides very good results and works correctly in case of uneven illumination.

The next steps are a few simple context filtering operations used to remove small objects. At first, dilation with a $3 \times 3$ kernel is applied. Then median filtering with a $5 \times 5$ window is performed, which helps to remove single outliers. Finally, erosion with a $3 \times 3$ kernel is used in order to keep the original dimensions of the remaining objects. An example result obtained after performing these filtrations is shown in Fig. 6.

The next step is the connected component labelling. An example result is shown in Fig. 7, where different objects are represented by different colours. As an output: area, coordinates of the bounding box and the centroid for each object are obtained. These values are used to calculate specific parameters, such as the size of the bounding box, its shape or the ratio of pixels belonging to the object to the total number of pixels inside it. The analysis of these parameters allowed us to determine a set of conditions that distinguishes the circles, squares and rectangles from other objects. For more details please refer to the provided source code [9]. In addition, the expected sizes of each of the mentioned figures are determined using altitude data provided by the used LiDAR, as their dimensions in pixels

**Fig. 6.** Sample image after filtration

**Fig. 7.** Sample result of the CCL module

depend on the distance from the camera. The main motivation for this approach was to reduce the number of false detections.

To accurately detect the entire marker, not just individual shapes, we performed a detailed analysis of the bounding boxes. The square, the rectangle and the small circle are considered as correctly detected if their bounding boxes are inside the bounding box of the large circle. This approach has significantly reduced the number of false detections.

Finally, the location and orientation of the detected marker is determined (cf. Fig. 8). The centroids of the square and rectangle are used to calculate the orientation. Firstly, the distance between these points is determined and compared to the diameter of the large circle to avoid incorrect detections. If that distance is in the appropriate range, the ratio of the difference between them along the Y axis to the corresponding difference along the X axis is specified. Values calculated in this way enable to determine the angle using the arctan function.

Then the centroids of the square and the rectangle are used to calculate the position of the marker on the image. The average values of the two mentioned centroids proved to be more reliable than the centroid of the big circle, especially in the case of an incomplete marker or low altitude. However, the prior detection of the big circle is crucial as the squares and rectangles were analysed only inside it. That means all three figures are necessary to estimate the position and orientation of the marker. At low altitude, when the big circle is not entirely visible, the centroid of the small circle is used as the location of the landing pad centre. The analysis of the image shown in Fig. 8 was used in the example below to calculate the position and orientation of the drone relative to the landing pad. The following results were obtained:

– horizontal distance from the centre of the image: $+40$ pixels
– vertical distance from the centre of the image: $+21$ pixels
– deviation from the fixed marker orientation: $-49°$

After taking into account the current drone altitude and known dimensions of the marker, the calculated values were converted into centimetres. The following results are obtained:

– horizontal offset from the centre of the image: 6.2 cm
– vertical offset from the centre of the image: 3.3 cm

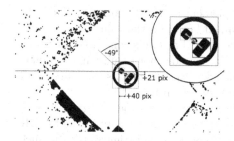 

**Fig. 8.** Sample result of the proposed vision algorithm. The centre of the camera field of view is indicated by two crossing magenta lines. The bounding box surrounding the circle is marked in red, the square in green and the rectangle in blue. The red dot marks the centre of the marker and the orange line marks its orientation. The small circle is not marked by its bounding box, because it is too small to be detected correctly. In the top right corner of the image the marker is enlarged for better visualization. The obtained drone position and orientation are also presented. (Color figure online)

**Fig. 9.** The working hardware-software system. The image is processed by the Arty Z7 board on the drone at a altitude of 0.5 m and the detected rectangle (blue bounding box) and square (green bounding box) are displayed on the monitor. (Color figure online)

## 4 HW/SW Implementation

We implemented all necessary components of the described algorithm in a heterogeneous system on the Arty Z7 development board with the Zynq SoC device. All image processing and analysis stages are implemented in the programmable logic of the SoC. It receives and decodes consecutive frames from the camera through a HDMI interface in native mode and performs the required image preprocessing operations in a pipelined way with IP cores connected in an established order. The performed operations in programmable logic are: conversion from RGB to greyscale, Gaussian low-pass filtration, median filtering, erosion, dilation and two more complex image analysis algorithms: connected component labelling (CCL) from our previous work [2] and adaptive thresholding. Finally, it sends the results to the processing system (the ARM processor) via the AXI interface. Moreover, in the prototyping phase the image processing results were transmitted via HDMI and visualized on a temporarily connected LCD screen.

The adaptive thresholding algorithm, mentioned in Sect. 3.3, consists of two stages. The first one finds the minimum and maximum values in $128 \times 128$ windows and calculates the appropriate thresholds, according to Eq. 1. The raw video stream does not contain pixel coordinates, but only pixel data and horizontal and vertical synchronization impulses, which is why the coordinates need to be calculated in additional counters implemented in programmable logic. In the second stage, local thresholds are obtained based on the values in the adjacent windows (4, 2 or 1). It should be noted that thresholds computed for frame $N-1$, are used on frame $N$. This approach, which does not require image buffering, is justified by the high sampling frequency of the camera as there are small differences in brightness of pixels on the subsequent frames.

In addition, we use the programmable logic part to supervise the altitude measurement using LiDAR. We use a simple state machine to control the LiDAR in PWM (Pulse Width Modulation) mode. It continuously triggers the distance measurement and reads its result. We implemented all but one modules as separate IP cores in the Verilog hardware description language. The only component that we did not create was an HDMI decoding module, which was obtained from Digilent company library (board vendor).

For the processing system, i.e. the ARM Cortex-A9 processor, we developed an application in the C programming language. Its main task is to fuse data from the vision system and LiDAR and send appropriate commands to the Pixhawk controller via the MAVLink 2.0 protocol using UART. In particular, it filters the objects present in the image using data obtained from the CCL module, searches for proper geometric shapes and determines the position of the drone relative to the landing pad[1]. In addition, the application controls the radio module for communication with the ground station. Thanks to this, we can remotely supervise the algorithm. The application also uses interrupts available in the processing system, to correctly read all input data streams. In this manner, it can stop the algorithm execution at any time, especially when an unexpected event occurs.

The resource utilization of the programmable logic part of the system is as follows: LUT 4897 (28.00%), FF (21368 (20.08%), BRAM (24 (17.14%) and DSP 25 (11.36%). It can be concluded that it is possible to implement improvements to the algorithm or to add further functionalities to the system. The estimated power usage is 2.191 W. In terms of performance, the programmable logic part works in real-time for a $1280 \times 720$ @ 60 fps video stream. This requires a clock rate of 74.25 MHz. We performed a static timing analysis and it turned out that the CCL module is currently the performance bottleneck, as it works with a maximal clock of 96 MHz. Therefore, for higher video stream resolutions processing this module has to be redesigned. Another performance limitation is the currently used communication with the processor through AXI interface. However, if required, there are more efficient options, like DMA. A photo of the working system is presented in Fig. 9.

## 5   Evaluation

To evaluate the system, several test sequences were recorded using the Yi camera mounted on the drone. The sequences were recorded both indoor and outdoor to evaluate the performance of the algorithm in different conditions. The altitude

---

[1] This is the same code as used in the software model [9].

data from the LiDAR sensor was also stored. Diverse images of the landing pad, i.e. for drone altitude from 0 to 1.5 m, for different orientations of the marker, for its different position in the image and on different backgrounds (outside and inside) were obtained.

At first, the marker detection rate was determined. 75 images were selected from the sequences recorded in varying conditions (sunny, cloudy, rainy) with the landing pad in different views. Then they were processed by the proposed algorithm. The marker was correctly detected on 48 images, which gives a 96% accuracy. The number of incorrectly detected shapes was also analysed – not a single shape outside the marker area has been falsely identified on the analysed images.

Secondly, the marker centre estimation accuracy was evaluated. The position returned by the algorithm was compared with a reference one (selected manually). The differences for the horizontal and vertical axes were calculated. Then the obtained results were averaged separately for each axis. For the horizontal axis the deviation was 0.40 pixels, while for the vertical axis it was 0.69 pixels.

The analysis of the presented results allows to draw the following conclusions about the performance of the vision algorithm. The percentage of frames with a correctly detected marker is high. However, some situations turned out to be problematic. These were mainly images, in which the marker was far from the centre of the camera's field of view. Analysing the marker centre estimation result it can be concluded that this value has been determined with high accuracy (below 1 pixel), which enables precise drone navigation.

## 6  Conclusion

In this paper a hardware-software control system for the autonomous landing of a drone on a static landing pad was presented. The use of a Zynq SoC device allowed to obtain real-time processing of a $1280 \times 720$ @ 60 fps video stream. The estimated power utilization of the system is 2.191 W. The designed algorithm has an accuracy of 96%. Unfortunately, due to the prevailing epidemic and the associated restrictions, it was not possible to test the fully integrated system (i.e. to perform a fully autonomous landing based on the proposed vision algorithm). As both the vision part and the communication with Pixhawk are tested, the only missing experiment is the test of the integrated system. So this will be the first step of further work and an opportunity to gather a larger database of test sequences in various conditions.

The next step will be the landing on a mobile platform – moving slowly, quickly or imitating the conditions on water/sea (when the landing pad sways on waves). In the vision part, it is worth considering using a camera with an even larger viewing angle and evaluate how it affects the algorithm. In addition, the methods that allow to distinguish the marker from the background – like RTV (Relative Total Variation) from work [4] should be considered. Another option is to use an ArUco marker, although implementing its detection in a pipeline vision system seems to be a greater challenge. An even more complicated Moreover, adding a Kalman filter (KF) to the system should increase reliability if detection errors occur incidentally on some frames in the video sequence. Additionally, the fusion of a video and IMU (Inertial Measurement Unit) data (from the Pixhawk flight controller) should be considered.

**Acknowledgement.** The work was supported by the Dean grant for young researches (first, second and third author) and AGH project number 16.16.120.773. The authors would like to thank Mr. Jakub Kłosiński, who during his bachelor thesis started the research on landing spot detection and Mr. Miłosz Mach, who was the initial constructor of the used drone.

# References

1. Carrio, A., Sampedro Pérez, C., Rodríguez Ramos, A., Campoy, P.: A review of deep learning methods and applications for unmanned aerial vehicles. J. Sens. 1–13 (2017)
2. Ciarach, P., Kowalczyk, M., Przewlocka, D., Kryjak, T.: Real-time FPGA implementation of connected component labelling for a 4K video stream. In: Hochberger, C., Nelson, B., Koch, A., Woods, R., Diniz, P. (eds.) ARC 2019. LNCS, vol. 11444, pp. 165–180. Springer, Cham (2019). https://doi.org/10.1007/978-3-030-17227-5_13
3. Giakoumidis, N., Bak, J.U., Gómez, J.V., Llenga, A., Mavridis, N.: Pilot-scale development of a UAV-UGV hybrid with air-based UGV path planning. In: Proceedings of 10th International Conference on Frontiers of Information Technology (FIT 2012), pp. 204–208, December 2012
4. Huang, Y., Zhong, Y., Cheng, S., Ba, M.: Research on UAV's autonomous target landing with image and GPS under complex environment. In: 2019 International Conference on Communications, Information System and Computer Engineering (CISCE), pp. 97–102 (2019)
5. Jin, S., Zhang, J., Shen, L., Li, T.: On-board vision autonomous landing techniques for quadrotor: a survey. In: 35th Chinese Control Conference (CCC), pp. 10284–10289 (2016)
6. Lee, S., Shim, T., Kim, S., Park, J., Hong, K., Bang, H.: Vision-based autonomous landing of a multi-copter unmanned aerial vehicle using reinforcement learning. In: 2018 International Conference on Unmanned Aircraft Systems (ICUAS), pp. 108–114 (2018)
7. Patruno, C., Nitti, M., Petitti, A., Stella, E., D'Orazio, T.: A vision-based approach for unmanned aerial vehicle landing. J. Intell. Robot. Syst. **95**(2), 645–664 (2019)
8. Qiu, R., Miao, X., Zhuang, S., Jiang, H., Chen, J.: Design and implementation of an autonomous landing control system of unmanned aerial vehicle for power line inspection. In: 2017 Chinese Automation Congress (CAC), pp. 7428–7431 (2017)
9. Software model of the proposed application. https://github.com/vision-agh/drone_landing_static. Accessed 07 July 2020
10. Xu, C., Tang, Y., Liang, Z., Yin, H.: UAV autonomous landing algorithm based on machine vision. In: IEEE 4th Information Technology and Mechatronics Engineering Conference (ITOEC), pp. 824–829 (2018)

# Exploration of Ear Biometrics with Deep Learning

Aimee Booysens and Serestina Viriri[✉]

School of Mathematics, Statistics and Computer Sciences,
University of KwaZulu-Natal, Durban, South Africa
viriris@ukzn.ac.za

**Abstract.** Ear recognition has become a vital issue in image processing to identification and analysis for many geometric applications. This article reviews the source of ear modelling, details the algorithms, methods and processing steps and finally tracks the error and limitations for the input database for the final results obtain for ear identification. The commonly used machine-learning techniques used were Naïve Bayes, Decision Tree and K-Nearest Neighbor, which then compared to the classification technique of Deep Learning using Convolution Neural Networks. The results achieved in this article by the Deep Learning using Convolution Neural Network was 92.00% average ear identification rate for both left and right ear.

## 1 Introduction

Biometric technology is a topic that has had a growing interest over the years because there is an increasing need for security and authentication, among others. Recognition systems have centred mainly on biometric features, and these are faces or fingerprints. These traditional features are widely studied, and their behaviour understood, while other less known biometric features, such as ears, have the potential to become the better applications.

The ear has an advantage over the traditional biometric features, as they have a stable structure that does not change as a person ages. It is a known fact that the face changes continually based on expressions; this does not occur with ears. Also, the ears environment are always known as they located on the sides of the head. In contrast, facial recognition typically requires a controlled environment for accuracy; this type of situation is not always present. Lastly, the ear does not need proximity to achieve capture, whereas the traditional biometric features do, like the eyes and fingers. The qualities mentioned above make the ear a promising field to study for recognition.

It is an accepted fact that the shape and appearance of the ears are unique per individual and that they are of fixed form during the lifetime of a person. According to reports, the variation over time in a human ear is most noticeable from when a person is four-months-old to eight years old and after the age of 70 years-old. The ear growth that occurs between four-months-old to eight years

© Springer Nature Switzerland AG 2020
L. J. Chmielewski et al. (Eds.): ICCVG 2020, LNCS 12334, pp. 25–35, 2020.
https://doi.org/10.1007/978-3-030-59006-2_3

old is linear, after that it is constant until the person is around 70 years old. At this age, the ears begin to increases again [1]. While there are small changes that take place in the ear structure, these are restricted to the ear lobe and are not linear. Such predictability of the ear makes it an exciting realm for research as machine-learning can quickly identify the ear and obtain a result.

Machine-learning is taking over the world and is becoming part of everybody's life. One of the branches of machine learning is that of Deep Learning,[2]. This branch deals with different algorithms to the structure and function of the human brain with multiple layers of a neural network.

In this article, we will be taking a look at how commonly used machine-learning techniques compare to Deep Learning using Convolution Neural Networks (CNN) in the identification of the ear, and the left and right ear.

## 2   Related Work

Zhang and Mu, [3], investigated ways of detection ears from 2D profile images. They investigated if multiple-scale faster region-based Neural Networks would work to detect images automatically. Shortfalls of this article are the following are not taken into account: pose variation, occlusion and imaging conditions. Even with these shortfalls, the article managed to achieve for web images 98% detection accuracy rate for both ears. The other two databases have the same test done to them, and this was the collection of J2 of University of Notre Dame Biometrics which achieved 100% for ear detection rate and the University of Beira Interior Ear Database (UBEAR) reached 98.22% for the ear detection rate.

Galdámez et al. proposed a solution to do ear identification [4]. This solution involved a Convolution Neural Networks (CNN) for any image inputted into the system and will give an output of the ear. The results that obtained from the CNN compared to the results for the same dataset of other machine-learning techniques of Principal Component Analysis (PCA), Linear discriminant analysis (LDA) and Speeded-up robust feature (SURF). These tests are carried out on two different datasets Avila's Police School and Bisite Video Dataset. On the first dataset, the model for CNN achieved a detection accuracy rate of 84.82%. CNN compared to PCA, LDA and SURF, only achieved a detection accuracy rate of 66.37%, 68.36% and 76.75% respectfully. The second dataset achieved CNN achieved a detection accuracy rate of 40.12% and compared to PCA which 21.83% ear accuracy rate, LDA which 27.37% ear accuracy rate and SURF which achieved an ear accuracy rate of 30.68%. The reason that the second dataset achieved so low is that the images obtained from videos.

Multimodal biometric systems address numerous problems observed in single modal biometric systems which are proposed by Amirthalingam and Radhamani, [5]. The sophisticated methods employed to find the right combination of multiple biometric modalities and various level of fusion applied to get the best possible recognition result discussed in this article. The combination of face and ear modality suggested, and the proposed framework of the biometric system are given. In this article, it claims that multi-biometrics improves over a single system, and uncorrelated modalities used to achieve performance in the multimodal system. This system produced an average ear detection rate of 72%.

## 3    Methods and Techniques

This research's depiction of the ear biometrics model, Fig. 1.

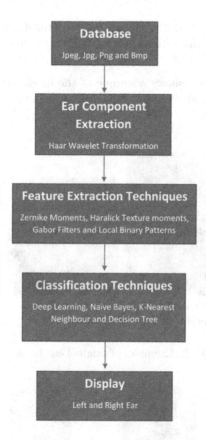

**Fig. 1.** Overview of ear identification system

### 3.1    Ear Components

The ear components are extracted using the Haar Wavelet Transformation (HWT) [6]. HWT is one of wavelet transformations. This transformation cross multiplies a function against the Haar Wavelet and is defined as Eq. (1).

$$yn = Hn.Xn \tag{1}$$

where $yn$ is the Haar Transformation at an $n$-input at function $Xn$. $Hn$ is the Haar Transformation which is a matrix which can be defined like Eq. (2).

$$\frac{1}{2}\begin{Bmatrix} 1 & 1 & \sqrt{2} & 0 \\ 1 & 1 & -\sqrt{2} & 0 \\ 1 & -1 & 0 & \sqrt{2} \\ 1 & -1 & 0 & -\sqrt{2} \end{Bmatrix} \tag{2}$$

Some of the properties that the Haar Wavelet Transformation [6] has is that there is no need for multiplications and that the input and output matrix of the same length. The components that were extracted are the left ear and right ear.

The figures in 2 and 3 shows a sample of the images that were used in this article and how the Haar Wavelet Transformation extracts the ear from the original facial or 2D image.

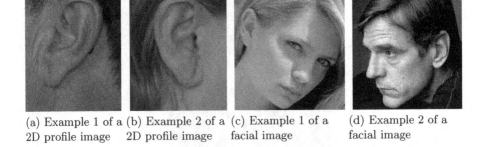

(a) Example 1 of a 2D profile image  (b) Example 2 of a 2D profile image  (c) Example 1 of a facial image  (d) Example 2 of a facial image

**Fig. 2.** Examples of original ear images

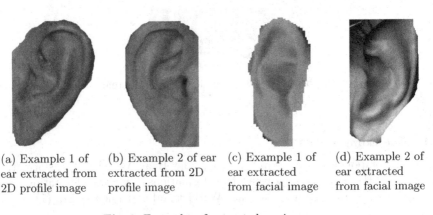

(a) Example 1 of ear extracted from 2D profile image  (b) Example 2 of ear extracted from 2D profile image  (c) Example 1 of ear extracted from facial image  (d) Example 2 of ear extracted from facial image

**Fig. 3.** Examples of extracted ear images

## 3.2   Feature Extraction

Feature extraction techniques are used on each ear image in the dataset to obtain a feature vector. Analysis was done on different textural, structural or geometrical feature extractions. The feature extraction techniques that were used are Zernike Moments, Local Binary Pattern (LBP), Gabor Filter and Haralick texture Moments. Investigation that was done in order to obtain the correct feature extraction for all component.

**The Zernike Moments** are used to overcome redundancy in which certain geometric moments obtain [7]. They are a class of orthogonal moments which are rotational invariant and effective in image representation. Zernike moments are a set of complex, orthogonal polynomials defined as the interior of the unit circle. The general form of the Zernike Moments defined in Eq. (3).

$$Z_{nm}(x,y) = Z_{nm}(p,\theta) = R_{nm}(p)e^{jm\theta} \tag{3}$$

where $x$, $y$, $p$ and $\theta$ correspond to Cartesian and Polar coordinates respectively, $n \in Z^{+}$ and $m \in Z$, constrained to $n - m$ even, $m \leq n$

$$R_{nm}(p) = \sum_{k=0}^{\frac{n-m}{2}} \frac{(-1)^k(n-k)!}{k!(\frac{n+m}{2}-k)!(\frac{n-m}{2}-k)!}p^{n-2k} \tag{4}$$

where $R_{nm}(p)$ is a radial polynomial and $k$ is the order.

**The Haralick Texture Moments** are texture features that can analyse the spatial distribution of the image's texture features [7] with different spatial positions and angles. This research computer four of these Haralick Texture Moments – Energy, Entropy, Correlation and Homogeneity.

Entropy is the reflection of the disorder and the complexity of the texture of the images. This is defined using Eq. (5).

$$Entropy = \sum_{ij} \hat{f}(i,j) \log \hat{f}(i,j) \tag{5}$$

where $\hat{f}(i,j)$ is the $[i,j]$ entry if the grey level value of image matrix and $i$ and $j$ are points on the image matrix.

Energy is the measure of the local homogeneity and is the opposite of Entropy. It shows the uniformity of the texture of the images and computed using the Eq. (6).

$$Energy = \sum_{ij} \hat{f}(i,j)^2 \tag{6}$$

where $\hat{f}(i,j)$ is the $[i,j]$ entry if the grey level value of image matrix and $i$ and $j$ are points on the image matrix.

Homogeneity is the reflection of "equaliness" of the images' textures and scale of local changes in the texture of the images. If this result is high, then there is no difference between the regions with regards to the texture of the images and is defined in Eq. (7).

$$Homogeneity = \sum_i \sum_j \frac{1}{1 + (i-j)^2} \hat{f}_{i,j} \tag{7}$$

where $\hat{f}(i,j)$ is the $[i,j]$ entry if the grey level value of image matrix and $i$ and $j$ are points in the image matrix.

Correlation is the consistency of the texture of the images and described using equation (8).

$$Correlation = \sum_{ij} \frac{(i - \mu_i)(j - \mu_j)\hat{f}(i,j)}{\sigma_i \sigma_j} \tag{8}$$

in which $\mu_j$, $\mu_i$, $\sigma_i$ and $\sigma_j$ are described as:

$$\mu_i = \sum_{i=1}^{n} \sum_{j=1}^{n} i \hat{f}(i,j) \quad \mu_j = \sum_{i=1}^{n} \sum_{j=1}^{n} j \hat{f}(i,j) \tag{9}$$

$$\sigma_i = \sqrt{\sum_{i=1}^{n} \sum_{j=1}^{n} (i - \mu_i)^2 \hat{f}(i,j)} \quad \sigma_j = \sqrt{\sum_{i=1}^{n} \sum_{j=1}^{n} (i - \mu_j)^2 \hat{f}(i,j)} \tag{10}$$

**The Gabor Filters** are geometric moments which are the product between an elliptical Gaussian and a sinusoidal, [8]. Gabor elementary function is the product of the pulse with a harmonic oscillation of frequency.

$$g(t) = \epsilon^{-\alpha^2 (t - t_0)^2} \epsilon^{-i2\pi(f - f_o) + \phi} \tag{11}$$

where $\alpha$ is the time duration of the Gaussian envelope, $t_0$ denotes the centroid, $f_0$ is the frequency of the sinusoidal and $\phi$ indicates the phase shift.

**Local Binary Pattern (LBP)** is a geometric moment operator that described the surrounding of the pixels by obtaining a bit code of a pixel [9], this is defined using Eq. (12).

$$C = \sum_{k=0}^{k=7} (2^k b_k) \tag{12}$$

$$b_k = \begin{cases} 1, \sum_{k=0}^{k=7} (t_k \geq C) \\ \\ 0, \sum_{k=0}^{k=7} (t_k < C) \end{cases}$$

where $t_k$ is the grayscale amount. $b_k$ is the binary variables between 1 and 0 and $C$ is a constant value of 0 and 1.

## 3.3   Classification

The main classification technique in this article is Deep Learning with the use of CNN, and these results are compared to numerous other classification techniques. These are different supervised and unsupervised machine-learning algorithms. The classification techniques used are Naïve Bayes, K-Nearest Neighbor and Decision Tree.

**Deep Learning with Convolutional Neural Network** (CNN) is a generalisation of feed forward neural networks to a sequence [2]. Given that a particular sequence has a certain number of inputs $(x_1, ...., x_t)$ this will then compute a sequence of outputs $(y_1, ...., y_t)$ which is done by using the below Eq. (15) and (14):

$$h_t = \sigma(W^{hx}x^t + W^{hh}h_{t-1}) \tag{13}$$

$$y_t = W^{yh}h_t \tag{14}$$

The RNN is easy to apply the map sequence when there is alignment between and input and an output.

**Decision Tree** uses recursive partitioning to separate the dataset by finding the best variable [10], and using the selected variable to split the data. Then using the entropy, defined in Eqs. (15) and (16), to calculate the difference that variable would make on the results if chosen. If the entropy is zero, then that variable is perfect to use, else a new variable needs to be selected.

$$H(D) = -\sum_{i=1}^{k} P(C_i|D) \log_k(P(C_i|D)) \tag{15}$$

where the entropy of a sample $D$ concerns the target variable of $k$ possible classes $C_i$.

$$P(C_i|D) = \frac{number\ of\ correct\ observation\ for\ that\ class}{total\ observation\ for\ that\ class} \tag{16}$$

where the probability of class $C_i$ in $D$ is obtained directly from the dataset.

**Naïve Bayes** classify an instance by assuming the presence or absence of a particular feature. Checks if it is unrelated to the presence or absence of another feature, given in the class variable [10,11]. Naïve Bayes calculation is done by using the probability for which it occurred, as defined in Eq. (17).

$$P(x_1, ........, x_n|y) = \frac{\prod_{i=1}^{n} P(y)P(x_n|y)}{P(x_1, ........, x_n)} \tag{17}$$

where case $y$ is a class value, attributes are $x_1, ........, x_n$ and $n$ is the sample size.

**K-Nearest Neighbor (KNN)** classifies by using a majority vote of its neighbours. The case is assigned to the class with the most common amongst its dataset. A distance function measures the KNN, for example, Euclidean, as defined in Eq. (18).

$$d = \sqrt{\sum_{i=1}^{n} (x_i - q_i)^2} \tag{18}$$

where $n$ is the size of the data, $x_i$ is an element in the dataset, and $q_i$ is a central point.

## 4   Results and Discussion

This research used a union of four different image databases. The total dataset contained 2997 facial and 2D profile images of approximately 360 subjects, which had been split the dataset into two groups of left and right ear. The datasets that are used were Annotated Web Ears (AWE) [12] and Annotated Web Ears additional database (AWEA) [12], AMI Ear (AMI) [13] and IIT Delhi Ear Database (IIT) [1].

The analysis completed was to obtain which feature extraction technique would achieve the most accurate True Positive Rate (TPR) for ear identification. The results were obtained by taking a combination of the feature vectors for each component and then classified using K-Nearest Neighbour to observe which combination of feature extraction technique achieved the highest True Positive Rate, results obtained are in Table 1 and Table 2.

Zernike Moments is a geometric feature extraction technique which is a useful feature extraction technique as it correctly obtains the edges of the ear. This technique is used in the normalised feature vector as it captures the shape and the proportion of the ear. Zernike Moments on its own achieved a TPR of 82.71% for the left ear and 84.49% for the right ear.

Local Binary Pattern is to a geometric feature extraction technique. This feature extraction technique correctly identified the inner lines of the ear. This feature extraction technique is used in the normalised feature vector as it can detect the shape of the ear. Local Binary Pattern on its own obtained a TPR of 71.8% for the left ear and 73.8% for right ear.

Haralick Texture is a texture extraction technique. It works well as a feature extraction texture as it picks up the course and the colour gradient of the ear. Haralick Texture achieved an average TPR of 99.7% for the left ear and 86.69% for the right ear.

Gabor filter was the worst achieving geometric feature of all the feature vector techniques. This feature vector technique works well with the other feature vector techniques which achieved a TPR of 92.87%. Whereas alone it only achieved a TPR of 66.32% for the left ear and 70.72% for the right ear.

Table 1 shows the average TPR of all the combinations of these feature extraction techniques. If there are more feature extraction techniques used, the better the results are. Hence this reason why all four feature extraction techniques need to be used to obtain the feature vector. This vector is then fused and normalised to get ear identification.

**Table 1.** Accuracy rates for the ears for the combination of feature extraction techniques

| Feature techniques | Percentage (%) |
|---|---|
| Gabor Filter and Zernike Moments and Haralick Texture and Local Binary Pattern | 92,87 |
| Gabor Filter and Zernike Moments and Local Binary Pattern | 90,80 |
| Gabor Filter and Zernike Moments and Haralick Texture | 91,75 |
| Gabor Filter and Haralick Texture and Local Binary Pattern | 90,45 |
| Gabor Filter and Zernike Moments and Local Binary Pattern | 87,56 |
| Zernike Moments and Haralick Texture and Local Binary Pattern | 88,46 |
| Zernike Moments and Haralick Texture | 86,78 |
| Zernike Moments and Local Binary Pattern | 88,96 |
| Haralick Texture and Local Binary Pattern | 87,86 |
| Gabor Filter and Zernike Moments | 85,42 |
| Gabor Filter and Haralick Texture | 83,26 |
| Gabor Filter and Local Binary Pattern | 82,69 |

**Table 2.** Accuracy rates per ear per feature extraction techniques

|  | Zernike Moments | LBP | Gabor Filter | Haralick Texture Moments |
|---|---|---|---|---|
| Left ear | 82.71% | 71.8% | 66.32% | 99.7% |
| Right ear | 84.49% | 73.8% | 70.72% | 86.69% |

## 4.1 Results for both Left and Right Ear

Several empirical experiments were carried out to investigate if Deep Learning with CNN is a better machine-learning algorithm to determine ear. Deep Learning with CNN was compared to the results of that of the more commonly used machine-learning algorithms(Decision Trees, Naïve Bayes and K-Nearest Neighbour). The testing was done by filling the training datasets with randomly chosen images from the original dataset and then tested with the different machine-learning algorithms and Deep Learning using CNN.

The tests showed that the K-Nearest Neighbour machine-learning algorithm achieved 60.2% average ear detection rate. The worst machine-learning algorithm was Decision Tree which achieved 53.4% ear accuracy identification rate which is a variation of 6.8% between the worst and best ear accuracy identification rate.

Deep Learning using CNN and this achieved 91% ear detection rate, it achieved a better result than that of K-Nearest Neighbour machine-learning algorithm because of this Deep Learning is a better machine-learning algorithm than the commonly used ones.

**Table 3.** Comparison of related works results to this research results for ear biometrics identification

|  | Left ear | Right ear | Both Ears |
|---|---|---|---|
| Zhang and Mu [3] | - | - | 98.74% |
| Galdamez et al. [4] | - | - | 62.74% |
| Amirthalingam and Radhamani [5] | - | - | 72% |
| **This research** | **95.36%** | **88.64%** | **92%** |

Table 3 shows the comparison between the results achieved by related works for ear identification and this research work for ear identification. As demonstrated, this research results obtained a lower True Positive Rate than that of Zhang and Mu [3], which could be as a result of the research using a larger dataset. Whereas a comparison to that of the other research works, this research achieved a higher True Positive Rate (Fig. 4).

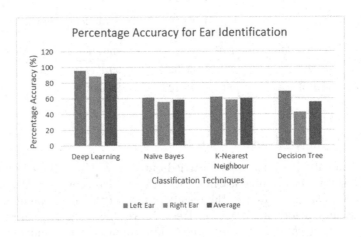

**Fig. 4.** Accuracy achieved for all classification techniques

# 5   Conclusion

This research presented an exploration of the ear biometric using Deep Learning Convolutional Neural Network. The ears were extracted and identified from 2D profile and facial images. The classification technique used was Deep Learning and was compared to more commonly used machine-learning algorithms. The identification that was done was left and right ear whereas other works only work on an average. The feature vector that was used is a combination of Haralick texture Moments, Zernike Moments, Gabor Filter and Local Binary Pattern. All these feature vectors are then fused and normalised to obtain a result. Naïve Bayes achieved 58.33% accuracy for the right ear, K-Nearest achieved 60.2% accuracy for the left ear, and Decision Tree achieved 55.72% accuracy for the left ear. Deep Learning achieved 95.36% accuracy for the left ear, 88.64% accuracy for the right ear. These results show that Deep Learning is a better way of obtaining results for the right and left ear. The total sum of the left and right ear identification rate of 92% was achieved.

# References

1. Kumar, A., Wu, C.: Automated human identification using ear imaging. Pattern Recogn. **45**(3), 956–968 (2012)
2. Schmidhuber, J.: Deep learning in neural networks: an overview. Neural Netw. **61**, 85–117 (2015)
3. Zhang, Y., Mu, Z.: Ear detection under uncontrolled conditions with multiple scale faster region-based convolutional neural networks. Symmetry **9**(4), 53 (2017)
4. Galdámez, P.L., Raveane, W., Arrieta, A.G.: A brief review of the ear recognition process using deep neural networks. J. Appl. Logic **24**, 62–70 (2017)
5. Amirthalingam, G., Radhamani, G.: A multimodal approach for face and ear biometric system. Int. J. Comput. Sci. Issues (IJCSI) **10**(5), 234 (2013)
6. Mulcahy, C.: Image compression using the Haar wavelet transform. Spelman Sci. Math. J. **1**(1), 22–31 (1997)
7. Teague, M.R.: Image analysis via the general theory of moments*. JOSA **70**(8), 920–930 (1980)
8. Berisha, S.: Image classification using Gabor filters and machine learning (2009)
9. Salah, S.H., Du, H., Al-Jawad, N.: Fusing local binary patterns with wavelet features for ethnicity identification. In: Proceedings of IEEE International Conference Signal Image Process, vol. 21, pp. 416–422 (2013)
10. Domingos, P.: A few useful things to know about machine learning. Commun. ACM **55**(10), 78–87 (2012)
11. Lowd, D., Domingos, P.: Naive bayes models for probability estimation. In: Proceedings of the 22nd International Conference on Machine Learning, pp. 529–536. ACM (2005)
12. Emersic, Z., Struc, V., Peer, P.: Ear recognition: more than a survey. Neurocomputing (2017)
13. Esther Gonzalez, L.A., Mazorra, L.: AMI Ear Database (2018). Accessed 3 Feb 2014

# Does Randomness in the Random Visual Cryptography Protocol Depend on the Period of Pseudorandom Number Generators?

Leszek J. Chmielewski$^{(\boxtimes)}$ (ID) and Arkadiusz Orłowski$^{(\boxtimes)}$ (ID)

Institute of Information Technology, Warsaw University of Life Sciences – SGGW,
Nowoursynowska 159, 02-775 Warsaw, Poland
{leszek_chmielewski,arkadiusz_orlowski}@sggw.edu.pl

**Abstract.** Actual randomness of the shares in the binary random visual cryptography protocol was tested with the NIST statistical test suite. Lack of randomness was noticed for a large secret image (shares 10K by 10K) in that the OverlappingTemplate test failed. The symptom of failure was much weaker for a twice smaller image and absent for a four times smaller one. The failure rate nonmonotonically depended on the period of the pseudorandom number generator, but was the largest for the worst generator. Four generators were used with periods from $2^{64}$ to $2^{19937}$. Hence, the reason for the nonrandomness seemed not to be the excessively large image size versus generator period, provided that the generator is reasonably good. Using the battery of tests made it possible to detect a gap in the randomness (not in the security) of the random visual cryptography protocol.

**Keywords:** Visual cryptography · Random shares · Randomness tests · Pseudorandom number generator · Period

## 1 Introduction

Transmitting an image in a secret way in the form of two or more images, called *shares*, where none of them contains any information on the transmitted secret, is an interesting and challenging target. Transmitting an image with the use of two truly random streams of bits has an additional advantage of hiding not only the contents of the secret, but also the mere fact that the information is transmitted. In this paper we shall present some results of extensive statistical randomness testing of the shares, in the cryptography protocol where the shares are intended to be fully random. The problem of our interest will be the relations of the randomness of shares, the period of pseudorandom number generators and the size of images. Testing the shares with the battery of randomness tests makes it possible to reveal even small gaps in the randomness of the coding process.

The history of black-and-white, binary visual cryptography with random shares seems to begin with the paper of Kafri and Keren [5], where the shares

© Springer Nature Switzerland AG 2020
L. J. Chmielewski et al. (Eds.): ICCVG 2020, LNCS 12334, pp. 36–47, 2020.
https://doi.org/10.1007/978-3-030-59006-2_4

are the same size as the secret image, and are truly random. After the decoding, the black pixel remains black, while the white one can be either white or black, with equal probability. In the protocol proposed by Naor and Shamir [8,9], which is considered the classic solution, the shares are twice as large as the secret. This expansion makes it possible to decode a black pixel as black 2×2 tile, and a white pixels a tile with exactly two white and two black pixels. Thus, the decoded image is darker than the secret, but it is very clear. The independence of the shares from the secret is maintained; however, their randomness is sacrificed. Numerous extensions of these protocols were proposed. To name just a few, let us note the coding of grey-level images [6] with multiple levels of access, color images [7], incremental decoding [18], or coding with shares which contain meaningful images not correlated with the secret [4,19]. The present state-of-the-art can be found, among others, in the book by Cimato and Yang [3], and in the review paper by Sharma, Dimri and Garg [14].

Recently, we proposed to use truly random shares in the protocol corresponding to the classic one [11] and we analyzed some compromise between its errors and randomness [12]. Extensions of this idea to color images are studied [2,10].

As a rule, in purely visual cryptography, where the secret image is revealed to a bare eye of the observer without any computations, the image has a reduced contrast and the details are partly blurred by imperfections emerging in the encryption-decryption process.

To our best knowledge, the actual randomness of the shares was not extensively tested with the suitable randomness tests, except some preliminary trials with the Runs test in [11] where single results were presented, and in [12] where series of random realizations were considered. In this paper we shall carry out the analysis with all the tests of the NIST statistical test suite [13]. We shall consider only the binary version of the method, and with one pair of shares.

The remainder of this paper is organized as follows. In the next Section the main elements of the classic visual cryptography protocol and the fully random visual cryptography protocol are briefly reminded. In Sect. 3 the application of the battery of 15 randomness tests from the NIST statistical test suite to the shares in the random protocol is described. In Sect. 4 the observation of the lack of randomness in one of the tests is presented, and its possible sources are commented on. In the last Section the paper is summarized and concluded.

## 2   Fully Random Visual Cryptography Protocol

In binary visual cryptography, the binary image called the *secret* is encoded in two images called the *shares*. Any share contains no information on the secret, but when the two shares are overlaid on one another they reveal the secret to the eye of the observer. This purely visual process called the *decoding* needs no equipment. In the case of binary cryptography, each pixel of the secret corresponds to a 2×2 pixels square, called the *tile*, in each share. In the coding process, one share called the *basic share* is generated randomly (according to some rules),

**Fig. 1.** All possible 2×2 tiles and their indexes. (Fig. 1 from [11].)

independently of the secret, and the other share called the *coding share* is the function of the first share and the secret. This is performed independently in each pixel of the secret, that is, in each pair (*basic tile, coding tile*) in the shares, respectively. All possible 2×2 tiles are shown in Fig. 1. Black color represents a black pixel, white color represents a transparent one.

In the classic Naor-Shamir protocol [8,9] as the basic tile one of the tiles having the equal numbers of black and white pixels is selected at random (tiles 4, 6, 7, 10, 11, 13 from Fig. 1). If the corresponding pixel of the secret is white, then as the coding tile, the same tile is taken. If this pixel is black, then the tile being the negative of the basic tile is taken as the coding tile. In this way, after overlaying the two shares on each other, the tile corresponding to a black pixel is fully black, and the tile corresponding to a white one is half-black and half-transparent, which makes an impression of a grey region when observed from a distance. For the secret shown in Fig. 2 further in the text this can be seen in the decoded image in Fig. 3a.

In the random protocol we proposed in [11], all the stages of processing each pixel of the secret are the same as in the classic protocol, but the basic tile is randomly drawn from among all the 16 possible tiles. This has two consequences. First, the whole basic and coding shares are random. Second, the decoded image has errors with respect to the classic scheme, where black becomes black and white becomes grey. This is due to that for some tiles it is impossible to represent a strictly half-white result for a white pixel of the secret. For example, if tile 2 (or 3, 5 or 9) is drawn in the basic share, then only one pixel can appear transparent in the reconstruction, instead of expected two; this is a −1 pixel error. If tile 1 is drawn, the number of transparent pixels will always be zero; this is a −2 pixel error. At the opposite, for tile 16 overlaid on itself, there will be 4 transparent pixels instead of 2, which is a +2 pixel error, etc. There are no errors in black pixels of the secret: a tile overlaid onto its negative always gives a black result. Restricting the set of possible tiles to avoid some of the errors results in some loss of randomness in the shares, which gives rise to three versions of the method [12]. In this paper we shall consider only the fully random version of the protocol, with all the 16 tiles, which we shall call simply the *random* one.

Randomness of the protocol is its valuable feature. Errors reduce the quality of the reconstructed image but it still remains readable. An example of decoding the image of Fig. 2 with the random method is shown in Fig. 3b, and the decoding errors are visualized in Fig. 4b further in the text.

# 3    Extensive Tests of Randomness

To approach the question of randomness more broadly than in [12], where only the Runs randomness test was used, here all the 15 tests from the NIST statistical test suite [13] are applied to the shares obtained from 100 random realizations of the coding. The results are shown in the form of histograms of the $p$-values. This illustrates not only the ratio of the count of tests which failed to the count of all tests, but also the distribution of $p$, so that its uniformity can be assessed visually. As parameters, the default values of the NIST software were used; in particular, $\alpha = 0.01$. Hence, failing tests were those for which $p \leq 0.01$.

To form the tested sequences, pixels were saved in two orderings: by rows and by columns. In this way, for each test there were four tested sequences, denoted, for the basic share, as 1h for pixels ordered horizontally (by rows) and 1v for pixels ordered vertically (by columns); for the coding share, the sequences were denoted 2h and 2v, respectively. We strived to show all the results for a single secret image in one page, in spite of that there are 15 tests, some with numerous subtests, four sequences for each. To have one graph for each test, we treated together the results for tests which have subtests. Hence, results for 188 tests, including their subtests, run 100 times for each of the four sequences, are shown in one figure, according to the concept of *small multiples* [17] which supports capturing the result at one glance.

The graphs are shown in Fig. 5 further in the text. The width of the histogram bins is $\alpha/4$. The counts for the failure of the tests, $p \leq \alpha$, denoted in the key of the graphs as *low*, are shown in shades of red. The histogram values for the success of the tests, denoted as *good*, are shown in the shades of grey. The data for the *basic shares*, denoted as share 1, are shown with full symbols, and for the *coding shares* – share 2, with empty symbols. The data for reading the pixels by columns, vertically, are shown with bars, and by rows, horizontally, with circles. Separately, to the right of the range of $p$, the number of cases for which the preconditions for the tests were not met are shown with the shades of blue and denoted with a question mark as *not applicable*. This can happen only for tests Universal, RandomExcursions and RandomExcursionsVariant.

To emphasize the interesting data which indicate the failures of the tests, $p \leq \alpha$, a nonlinear axis for $p$ is used. The transformation is $p \to p^r$, with $r$ such that the point 0.01 takes the place of the former point 0.1. In this way, the counts in four histogram bins for $p \leq \alpha$, each of width $\alpha/4$, can be clearly seen.

# 4    Lack of Randomness in One of the Tests

## 4.1    Observation

When performing the experiments, for one of the images we noticed that among the tests, the OverlappingTemplate gives results indicating weaker randomness than the other tests. This was the image of Fig. 2, version *Double* (size $3400 \times 2160$). The image is designed so that it is possible to see for what font size the decoding errors make the text hardly readable. The decoded versions of this

> ## \Large  14.4pt:  Praesent consequat, mi at iaculis iaculis, arcu turpis suscipit ante, nec auctor elit neque nec. . .
>
> \large 12pt: Praesent consequat, mi at iaculis iaculis, arcu turpis suscipit ante, nec auctor elit neque nec turpis. Curabitur sodales lorem dapibus, condimentum justo a, tincidunt nunc. In id sem varius,. . .
>
> \normalsize 10pt: Praesent consequat, mi at iaculis iaculis, arcu turpis suscipit ante, nec auctor elit neque nec turpis. Curabitur sodales lorem dapibus, condimentum justo a, tincidunt nunc. In id sem varius, malesuada tortor non,. . .
>
> \small=\footnotesize 9pt: Praesent consequat, mi at iaculis iaculis, arcu turpis suscipit ante, nec auctor elit neque nec turpis. Curabitur sodales lorem dapibus, condimentum justo a, tincidunt nunc. In id sem varius, malesuada tortor non, venenatis enim. Morbi tincidunt sapien nec lectus volutpat, non gravida orci suscipit. Duis tincidunt. . .
>
> \scriptsize 7pt: Praesent consequat, mi at iaculis iaculis, arcu turpis suscipit ante, nec auctor elit neque nec turpis. Curabitur sodales lorem dapibus, condimentum justo a, tincidunt nunc. In id sem varius, malesuada tortor non, venenatis enim. Morbi tincidunt sapien nec lectus volutpat, non gravida orci suscipit. Duis tincidunt mollis lobortis. Ut mauris odio, pharetra finibus ligula nec, vehicula iaculis ligula. Mauris. . .
>
> \tiny 5pt: Praesent consequat, mi at iaculis iaculis, arcu turpis suscipit ante, nec auctor elit neque nec turpis. Curabitur sodales lorem dapibus, condimentum justo a, tincidunt nunc. In id sem varius, malesuada tortor non, venenatis enim. Morbi tincidunt sapien nec lectus volutpat, non gravida orci suscipit. Duis tincidunt mollis lobortis. Ut mauris odio, pharetra finibus ligula nec, vehicula iaculis ligula. Mauris erat nunc, efficitur quis tincidunt at, eleifend nec dolor. Pellentesque eu metus ac ipsum tempus porttitor non eget diam. Vivamus a dui dapibus augue suscipit bibendum eu vitae ante. Aenean vitae pretium lectus. Quisque nisi purus, semper sed scelerisque nec, porta ac neque. Sed mollis tempor arcu, in luctus tellus fermentum id.

**Fig. 2.** The source binary image used in tests, size 1700×1080 (framed to show the area) – size Single. Sizes Double and Quadruple were formed by replacing each pixel with four ones, of the same value. This is a refined version of the image of Fig. 4 from [12].

**Table 1.** Abstracted results of all tests, for the image of Fig. 2 in three sizes. Numbers of failing, good and *not applicable* realizations and ratio of failures to all realizations shown as: upper part of each row – pairs (or triplets, where applicable) of numbers of realizations: (*low good n/a*); lower part of each row – ratio of *low* versus (*low+good*).

| # | Size | Dimensions | Frequency | Block Frequency | Cumulative Sums | Runs | LongestRun | Rank | FFT | Non Overlapping Template | Overlapping Template | Universal | Approximate Entropy | Random Excursions | Random Excursions Variant | Serial | Linear Complexity |
|---|---|---|---|---|---|---|---|---|---|---|---|---|---|---|---|---|---|
| 1 | Single | 1700×1080 | 0 400 | 2 398 | 0 800 | 2 398 | 3 397 | 4 396 | 4 396 | 573 58627 | 7 393 | 3 397 | 0 3 397 | 31 2761 408 | 46 6236 918 | 10 790 | 4 396 |
|   |   | ratio of *good* | 0.000 | 0.005 | 0.000 | 0.005 | 0.007 | 0.010 | 0.010 | 0.009 | **0.017** | 0.007 | 0.007 | 0.011 | 0.007 | 0.012 | 0.010 |
| 2 | Double | 3400×2160 | 0 400 | 8 392 | 0 800 | 5 395 | 6 394 | 3 397 | 4 396 | 590 58610 | 26 374 | 0 400 | 0 6 394 | 19 2957 224 | 50 6646 504 | 9 791 | 1 399 |
|   |   | ratio of *good* | 0.000 | 0.020 | 0.000 | 0.012 | 0.015 | 0.007 | 0.010 | 0.009 | **0.065** | 0.000 | 0.015 | 0.006 | 0.007 | 0.011 | 0.002 |
| 3 | Quadruple | 6800×4320 | 6 394 | 4 396 | 10 790 | 8 392 | 8 392 | 1 399 | 2 398 | 591 58609 | 175 225 | 6 394 | 0 2 398 | 32 3056 112 | 72 6876 252 | 5 795 | 4 396 |
|   |   | ratio of *good* | 0.015 | 0.010 | 0.012 | 0.020 | 0.020 | 0.002 | 0.005 | 0.009 | **0.437** | 0.015 | 0.005 | 0.010 | 0.010 | 0.006 | 0.010 |

image in three sizes are shown in Fig. 3, and the decoding errors in Fig. 4. The histograms of $p$ for the size *Double* are shown in Fig. 5. A non-uniformity of the histogram for OverlappingTemplate test, with some domination of counts for failures, can be seen only for this test. In brief, only the numbers of failures, successes and not applicable results for this image are collected in Table 1 row 2. The OverlappingTemplate test seems to be sensitive to some specific property of the sequence tested. This was the trail by which we expected to spot the source of the lack of randomness in some cases.

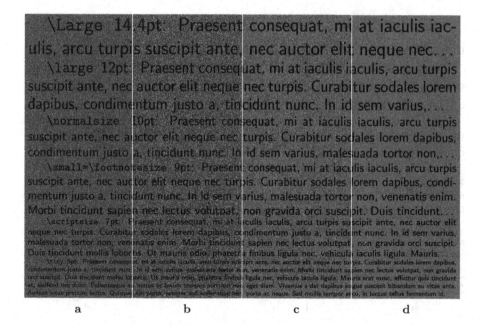

Fig. 3. Image of Fig. 2 decoded: (a) classic method, size Single; (b) random method, size Single; (c) random method, size Double; (d) random method, size Quadruple.

### 4.2   Hypothesis and Expectations

We hypothesized that this deficiency in randomness can have its source in the insufficient quality of the random (pseudorandom) number generator, namely the limited length of its period, in relation to the size of the coded image.

To check this, we considered four of the generators available in Matlab [15], with significantly different periods: shift-register generator summed with linear congruential generator, period $2^{64}$; multiplicative lagged Fibonacci generator, period $2^{124}$; modified subtract with borrow generator, period $2^{1492}$, and Mersenne twister, period $2^{19937}$ (see [16] and the references therefrom; lengths of the periods are approximate). We made calculations with these four generators and for three sizes (resolutions) of the same image. We expected the shorter the period and the larger the image, the larger the ratio of random realizations which fail in the OverlappingTemplate test.

### 4.3   Results and Discussion

The results are shown in Fig. 6: in a 3D graph with logarithmic scale to show the overall tendency, and in a series of 2D graphs with linear scale to visualize the changes more clearly. The histograms of $p$ from which the data for these graphs were taken are shown in Fig. 7.

The histograms for the smallest image exhibit very small or no signs of the lack of randomness (count for $p \leq 0.01$ equals 0.017, not much more than 0.01).

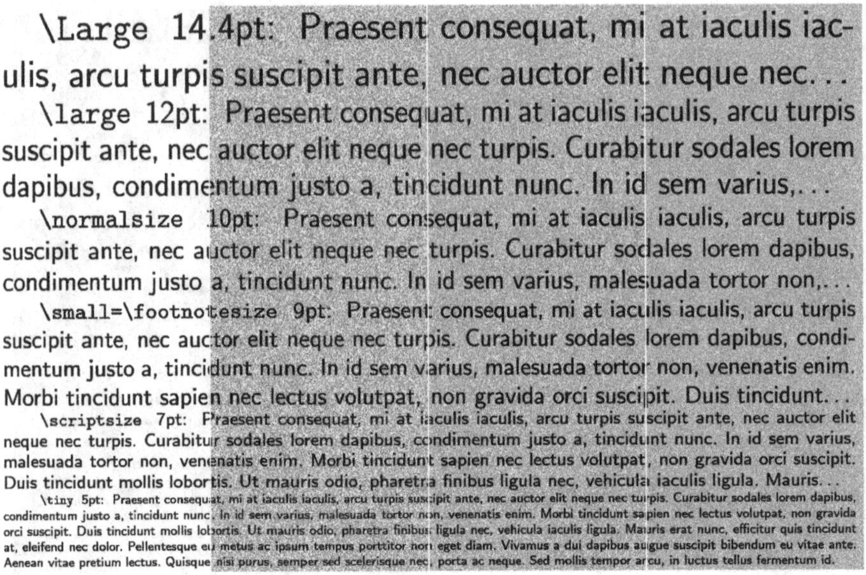

\Large 14.4pt: Praesent consequat, mi at iaculis iaculis, arcu turpis suscipit ante, nec auctor elit neque nec. . .

\large 12pt: Praesent consequat, mi at iaculis iaculis, arcu turpis suscipit ante, nec auctor elit neque nec turpis. Curabitur sodales lorem dapibus, condimentum justo a, tincidunt nunc. In id sem varius,. . .

\normalsize 10pt: Praesent consequat, mi at iaculis iaculis, arcu turpis suscipit ante, nec auctor elit neque nec turpis. Curabitur sodales lorem dapibus, condimentum justo a, tincidunt nunc. In id sem varius, malesuada tortor non,. . .

\small=\footnotesize 9pt: Praesent consequat, mi at iaculis iaculis, arcu turpis suscipit ante, nec auctor elit neque nec turpis. Curabitur sodales lorem dapibus, condimentum justo a, tincidunt nunc. In id sem varius, malesuada tortor non, venenatis enim. Morbi tincidunt sapien nec lectus volutpat, non gravida orci suscipit. Duis tincidunt. . .

\scriptsize 7pt: Praesent consequat, mi at iaculis iaculis, arcu turpis suscipit ante, nec auctor elit neque nec turpis. Curabitur sodales lorem dapibus, condimentum justo a, tincidunt nunc. In id sem varius, malesuada tortor non, venenatis enim. Morbi tincidunt sapien nec lectus volutpat, non gravida orci suscipit. Duis tincidunt mollis lobortis. Ut mauris odio, pharetra finibus ligula nec, vehicula iaculis ligula. Mauris. . .

\tiny 5pt: Praesent consequat, mi at iaculis iaculis, arcu turpis suscipit ante, nec auctor elit neque nec turpis. Curabitur sodales lorem dapibus, condimentum justo a, tincidunt nunc. In id sem varius, malesuada tortor non, venenatis enim. Morbi tincidunt sapien nec lectus volutpat, non gravida orci suscipit. Duis tincidunt mollis lobortis. Ut mauris odio, pharetra finibus ligula nec, vehicula iaculis ligula. Mauris erat nunc, efficitur quis tincidunt at, eleifend nec dolor. Pellentesque eu metus ac ipsum tempus porttitor non eget diam. Vivamus a dui dapibus augue suscipit bibendum eu vitae ante. Aenean vitae pretium lectus. Quisque nisi purus, semper sed scelerisque nec, porta ac neque. Sed mollis tempor arcu, in luctus tellus fermentum id.

a        b        c        d

**Fig. 4.** Decoding errors of images of Fig. 3: (**a**) classic method: no errors; (**b**) random method, size Single; (**c**) random method, size Double; (**d**) random method, size Quadruple. Errors shown with colors: dark red (■): $-2$ pix error; beige (■): $-1$ pix error; turquoise (■): $+1$ pix error; teal (■): $+2$ pix error (palette designed in [11] with ColorBrewer [1]). (Color figure online)

For the medium-sized image the counts for $p \leq 0.01$ are generally larger than those for other values of $p$. This phenomenon is very pronounced for the largest image. However, the dependence of the phenomenon on the period of the generator is not fully univocal. There is a slight decrease of failure ratios with the growing generator period for smaller images. For the large one the changes are not monotonic, but evidently for the smallest period the failure ratio is the largest. The differences observed are small so their significance can be doubted. Consequently, the expected phenomenon appeared to some extent, but its intensity was weak. Simultaneously, the counts strongly increase in the direction of the image size. No prevailing pattern can be noticed concerning in which of the shares – basic or coding, and in which direction – vertical or horizontal, the ratio of failures is the largest.

### 4.4 Conclusions from the Experiment and Comments

From this experiment two conclusions can be drawn. First, the quality of the generator, measured by its period, is not important in the tested case, provided that a sufficiently good generator is used. Second, the dependence on the image size is strong. Below we attempt to comment on these results.

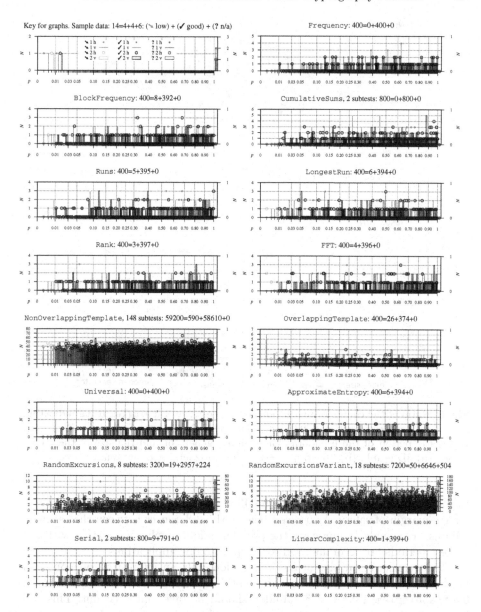

**Fig. 5.** Histograms of $p$ for image `TextBlack` in double size. Key shown in the upper left sub-image.

What concerns the quality of the generator, it seems that any reasonably good generator should be enough, due to that there are a very limited number of tiles in the set of tiles used to code the secret image. The result for the generator with the period $2^{64}$ was the worst besides that still the relation of this period

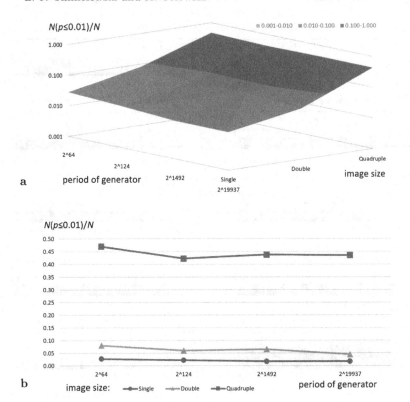

**Fig. 6.** Ratio of the counts of realizations of the test `OverlappingTemplate` indicating lack of randomness (with $p \leq 0.01$) in relation to the counts of all the realizations, versus the period of the pseudorandom number generator and the size of the coded image. (**a**) 3D graph; (**b**) 2D graphs. Scales for horizontal axes are qualitative.

to the size of each share is $2^{64}/(6800 \times 4320 \times 4) \approx 1.57 \times 10^{11}$ which constitutes a huge margin of safety. As noticed before, the differences between the ratios of failures were small. In any case, the pseudorandom number generators with the longest periods are recommended for cryptographic applications.

What concerns the size of the image, definitely an interesting phenomenon has been observed. To carry out its in-depth analysis, the principle on which the `OverlappingTemplate` test is founded should be taken into account. This can be the starting point for a further study.

It is interesting that in the largest image the amount of nonrandomness, measured with the ratio of failures, is similar in the basic shares (denoted with no. 1 and with full symbols) as well as in the coding shares (no. 2, empty symbols). The basic share is generated without any reference to the coded secret. This could suggest that the way it is generated has some deficiency, irrespective of the generator used. The only specific element of this process is that the basic share emerges as a superposition of tiles, not as a directly generated image. It can be conjectured that the direct generation of a random sequence of pixels

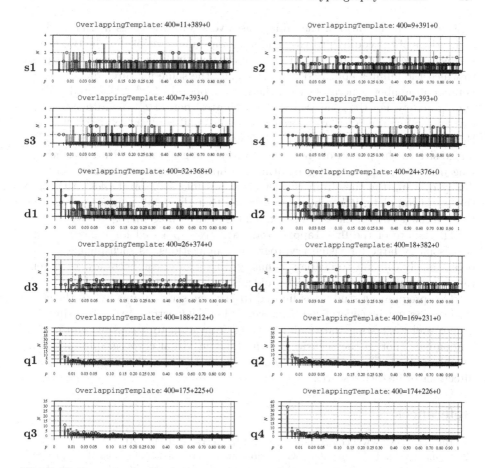

**Fig. 7.** Histograms of $p$ for the test `OverlappingTemplate`, image `TextBlack` in three sizes, and with four random number generators. Image sizes: (**sN**) single, (**dN**) double, (**qN**) quadruple. Periods of random number generators: (**r1**) $2^{64}$, (**r2**) $2^{124}$, (**r3**) $2^{1492}$, (**r4**) $2^{19937}$. Key shown in the upper left sub-image of Fig. 5.

could make a difference. However, equally as it is in the present algorithm, the so generated share would have tiles belonging to the set of 16 possible tiles shown in Fig. 1, displaced randomly. Nevertheless, this hypothesis is worth verifying.

## 5   Summary and Conclusion

A study on the randomness of the shares in the random visual cryptography protocol was conducted with the use of the tests contained in the NIST statistical test suite. A trace of the lack of randomness was noticed in some cases.

It was observed that for a binary image sized around $5K \times 5K$ pixels, for which the shares are around $10K \times 10K$ pixels large, the shares did not pass the `OverlappingTemplate` test from the NIST test suite. This phenomenon was measured with the ratio of the failing cases versus all the cases in 100 random realizations of the coding process. The same was faintly visible for the two times smaller image and did not happen at all for the four times smaller one. The extent to which the failures appeared was practically the same for three of four pseudorandom number generators with the longest periods: $2^{124}$, $2^{1492}$, and $2^{19937}$, approximately. For the generator with the shortest period $2^{64}$ the rate of failures was larger, but the relative difference was too small to draw solid conclusions. Hence, the phenomenon did not result from an excessively large quotient of the image size to the period of the generator, for the generator with a reasonably long period. The process of generating the first, coding share, consists in randomly drawing the tiles from the set of 16 possible $2 \times 2$ pixel tiles. Although this set is complete, it has been conjectured that the small number of possible tiles can be a bottleneck for the randomness of the whole share.

The failure of the `OverlappingTemplate` test appeared for images of size unreasonably large for visual cryptography applications, as such images would be extremely difficult to overlay manually with a sub-pixel accuracy. However, this failure can point to potential problems with randomness in general.

This preliminary study was limited to just one series of images, so the conclusions can not be generalized. However, it has demonstrated that by using the proper tool for testing the randomness, an indication to a possible gap in the randomness (not in the security) of the random visual cryptography protocol can be spotted. The observations made can be the cues showing the possible directions of further research.

# References

1. Brewer, C.A.: www.colorbrewer.org (2018). Accessed 9 May 2020
2. Chmielewski, L.J., Gawdzik, G., Orłowski, A.: Towards color visual cryptography with completely random shares. In: Proceedings of the Conference PP-RAI 2019 - Konf. Polskiego Porozumienia na rzecz Rozwoju Sztucznej Inteligencji, pp. 150–155. Wrocław University of Science and Technology, Wrocław, Poland, 16–18 October 2019. http://pp-rai.pwr.edu.pl/PPRAI19_proceedings.pdf
3. Cimato, S., Yang, C.N.: Visual Cryptography and Secret Image Sharing, 1st edn. CRC Press Inc., Boca Raton (2017). https://www.crcpress.com/Visual-Cryptography-and-Secret-Image-Sharing/Cimato-Yang/p/book/9781138076044
4. Dhiman, K., Kasana, S.S.: Extended visual cryptography techniques for true color images. Comput. Electr. Eng. **70**, 647–658 (2018). https://doi.org/10.1016/j.compeleceng.2017.09.017
5. Kafri, O., Keren, E.: Encryption of pictures and shapes by random grids. Opt. Lett. **12**(6), 377–379 (1987). https://doi.org/10.1364/OL.12.000377
6. Lin, C.C., Tsai, W.H.: Visual cryptography for gray-level images by dithering techniques. Pattern Recogn. Lett. **24**(1), 349–358 (2003). https://doi.org/10.1016/S0167-8655(02)00259-3
7. Liu, F., Wu, C.K., Lin, X.J.: Colour visual cryptography schemes. IET Inf. Secur. **2**(4), 151–165 (2008). https://doi.org/10.1049/iet-ifs:20080066
8. Naor, M., Shamir, A.: Visual cryptography. In: De Santis, A. (ed.) EUROCRYPT 1994. LNCS, vol. 950, pp. 1–12. Springer, Heidelberg (1995). https://doi.org/10.1007/BFb0053419

9. Naor, M., Shamir, A.: Visual cryptography II: improving the contrast via the cover base. In: Lomas, M. (ed.) Security Protocols. LNCS, vol. 1189, pp. 197–202. Springer, Heidelberg (1997). https://doi.org/10.1007/3-540-62494-5_18

10. Orłowski, A., Chmielewski, L.J.: Color visual cryptography with completely randomly coded colors. In: Vento, M., Percannella, G. (eds.) CAIP 2019. LNCS, vol. 11678, pp. 589–599. Springer, Heidelberg (2019). https://doi.org/10.1007/978-3-030-29888-3_48

11. Orłowski, A., Chmielewski, L.J.: Generalized visual cryptography scheme with completely random shares. In: Petkov, N., Strisciuglio, N., Travieso, C.M. (eds.) Proc. 2nd International Conference Applications of Intelligent Systems APPIS 2019, pp. 33:1–33:6. Association for Computing Machinery, Las Palmas de Gran Canaria (2019). https://doi.org/10.1145/3309772.3309805

12. Orłowski, A., Chmielewski, L.J.: Randomness of shares versus quality of secret reconstruction in black-and-white visual cryptography. In: Rutkowski, L., et al. (eds.) ICAISC 2019. LNAI, vol. 11509, pp. 58–69. Springer, Cham (2019). https://doi.org/10.1007/978-3-030-20915-5_6

13. Rukhin, A., Soto, J., Nechvatal, J., et al.: A statistical test suite for random and pseudorandom number generators for cryptographic applications. Technical report 800–22 Rev 1a, National Institute of Standards and Technology, 16 September 2010. http://www.nist.gov/manuscript-publication-search.cfm?pub_id=906762, series: Special Publication (NIST SP)

14. Sharma, R.G., Dimri, P., Garg, H.: Visual cryptographic techniques for secret image sharing: a review. Inf. Secur. J.: A Glob. Perspect. **27**(5–6), 241–259 (2018). https://doi.org/10.1080/19393555.2019.1567872

15. The MathWorks Inc: MATLAB. Natick, MA, USA (2020). https://www.mathworks.com

16. The MathWorks Inc: MATLAB documentation: Creating and Controlling a Random Number Stream. Natick, MA, USA (2020). https://www.mathworks.com/help/matlab/math/creating-and-controlling-a-random-number-stream.html. Accessed 04 Jan 2020

17. Tufte, E.R.: The Visual Display of Quantitative Information. Graphic Press, Cheshire (1983)

18. Wang, R.Z., Lan, Y.C., Lee, Y.K., et al.: Incrementing visual cryptography using random grids. Opt. Commun. **283**(21), 4242–4249 (2010). https://doi.org/10.1016/j.optcom.2010.06.042

19. Wu, H., Wang, H., Yu, R.: Color visual cryptography scheme using meaningful shares. In: Proceedings of the 8th International Conference Intelligent Systems Design and Applications ISDA 2008, vol. 3, pp. 173–178. Kaohsiung, Taiwan, November 2008. https://doi.org/10.1109/ISDA.2008.130

# Software for CT-image Analysis to Assist the Choice of Mechanical-Ventilation Settings in Acute Respiratory Distress Syndrome

Eduardo Enrique Dávila Serrano[1], François Dhelft[1,2], Laurent Bitker[1,2], Jean-Christophe Richard[1,2], and Maciej Orkisz[1(✉)]

[1] Univ Lyon, Université Claude Bernard Lyon 1, INSA-Lyon, UJM-Saint Etienne, CNRS, Inserm, CREATIS UMR 5220, U1206, 69621 Lyon, France
{davila,jean-christophe.richard,maciej.orkisz}@creatis.insa-lyon.fr
[2] Service de Réanimation Médicale,
Hôpital de la Croix Rousse, Hospices Civils de Lyon, Lyon, France
{francois.dhelft,laurent.bitker,j-christophe.richard}@chu-lyon.fr,
https://www.creatis.insa-lyon.fr/site7/en

**Abstract.** Acute respiratory distress syndrome (ARDS) is a critical impairment of the lung function, which occurs – among others – in severe cases of patients with Covid-19. Its therapeutic management is based on mechanical ventilation, but this may aggravate the patient's condition if the settings are not adapted to the actual lung state. Computed tomography images allow for assessing the lung ventilation with fine spatial resolution, but their quantitative analysis is hampered by the contrast loss due to the disease. This article describes software developed to assist the clinicians in this analysis by implementing semi-automatic algorithms as well as interactive tools. The focus is the assessment of the cyclic hyperinflation, which may lead to ventilator-induced lung injury. For this purpose aerated parts of the lungs were segmented in twenty ARDS patients, half with Covid-19. The results were in very good agreement with manual segmentation performed by experts: 5.3% (5.1 ml) mean difference in measured cyclic hyperinflation.

**Keywords:** Acute respiratory distress syndrome · Hyperinflation · Lung segmentation. · Computed tomography · Thoracic images

## 1 Introduction

Acute respiratory distress syndrome (ARDS) is a particularly severe impairment of the lung function resulting in a very high mortality. Its therapeutic management is based on mechanical ventilation with positive end-expiratory pressure

This work was performed within the framework of the LABEX PRIMES (ANR-11-LABX-0063) of Université de Lyon, within the program "Investissements d'Avenir" (ANR-11-IDEX-0007) operated by the French National Research Agency (ANR).

L. J. Chmielewski et al. (Eds.): ICCVG 2020, LNCS 12334, pp. 48–58, 2020.
https://doi.org/10.1007/978-3-030-59006-2_5

(PEEP) used in an attempt to control hypoxemia and hypercapnia, and thus keep the patient alive during the treatment of the syndrome's actual cause. Nevertheless, experimental and clinical studies have repeatedly demonstrated that inadequate ventilator settings are likely to aggravate ARDS lung injury [9]. Customizing these settings requires quantitative assessment of each patient's lung response to mechanical ventilation.

Thoracic computed tomography (CT) is the reference technique for quantifying pulmonary aeration in vivo [2] provided that the lungs are delineated from the surrounding tissues. Unfortunately, while segmenting healthy lungs in CT scans is eased by high contrast between the lung parenchyma and other tissues, the segmentation of diseased lungs remains challenging [7]. In particular, the radiological hallmark of ARDS is the presence of large heterogeneous non-aerated (opaque) regions (Fig. 1), which defeat the existing segmentation methods and require human interaction to improve results. Literature on automatic lung segmentation for ARDS is very limited. Recently published methods obtained promising results, but were designed for animal models of the syndrome [4,8]. For these reasons, as well as due to the lack of dedicated software, CT is still not used in clinical practice to assist the physicians in setting the main ventilation parameters: PEEP and tidal volume (VT).

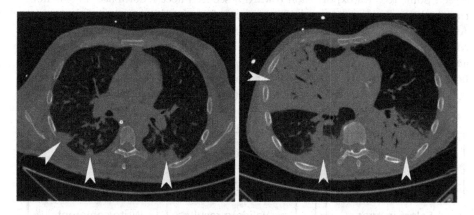

**Fig. 1.** Examples of slices from thoracic CT scans of patients with mild (left) and severe (right) ARDS. Arrows indicate opaque regions that cannot be distinguished from the surrounding tissues based on gray-levels

To address the clinicians' needs we developed a software designed for quantitative analysis of CT-scan pairs acquired at end-inspiration (referred to as *inspi*) and end-expiration (*expi*). The project described in this article specifically tackles the assessment of cyclic hyperinflation, i.e., locally excessive lung inflation at each insufflation from the ventilator, which probably reflects too large tidal volume and may lead to ventilator-induced lung injury (VILI). This task requires the segmentation of two three-dimensional (3D) CT images, each of the order of

$512^3$ voxels with sub-millimeter spatial resolution. To be usable in the clinical practice, the whole process – from image loading to numerical result display – must hold within a few minutes.

To achieve these objectives, a dedicated graphical user interface (GUI) implements a series of relatively simple algorithms, mainly based on seeded region growing and mathematical morphology, as well as smart interactive tools allowing the user to quickly edit the result if necessary. Although none of these components is novel, their assembly constitutes a new software that fills the void and finally makes it possible to provide decision support within a timeframe compatible with clinical practice. Herein, we will first describe the main components, then preliminary results on twenty ARDS patients will be reported.

## 2    Method

Figure 2 displays the main functions available via the GUI of the software, which is helpful to summarize the workflow. The left panel encapsulates *inspi* and *expi* image anonymization, loading, and visual inspection, as well as manual entry of data missing in the DICOM header, such as PEEP, VT, and patient's height. The right panel contains the whole processing pipeline from an interactive selection of a seed point to data exportation toward a secured-access database. Let us note, that each step can be repeated, if necessary, after visual inspection of the results. As the data can be pushed onto the database at any stage of the process, the completeness of the process needs to be controlled via the summary that gives access to a table specifying the already performed steps with their date and hour, as well as to the database itself via a web-page. The subsequent sections will describe in details the segmentation and interactive-correction steps.

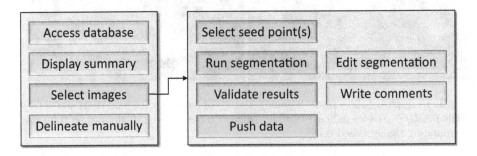

**Fig. 2.** Main functions available via the graphical user interface. Mandatory operations are highlighted by color. **Run segmentation** and **Edit segmentation** steps are respectively described in Sects. 2.1 and 2.2 (Color figure online)

## 2.1  Segmentation

As this work deals with the aerated part of the lungs, like in [5], let us specify the relationship between aeration and CT-image gray levels, actually CT-numbers in Hounsfield units (HU). This relationship is almost linear for lung tissues, where $I_{air} \leq I(\mathbf{x}) \leq I_{water}$, so that air fraction in a voxel $\mathbf{x}$ is calculated as [10]:

$$a(\mathbf{x}) = \frac{I_{water} - I(\mathbf{x})}{I_{water} - I_{air}},\qquad(1)$$

with $I_{air} = -1000\,\mathrm{HU}$ corresponding to pure air and $I_{water} = 0\,\mathrm{HU}$ corresponding to air-free parenchymal tissue (alveolar walls, capillary blood vessels, etc.). In normal parenchyma, air fraction is comprised between 50 and 90%, i.e., $I(\mathbf{x}) \in [-900, -500)\,\mathrm{HU}$. Above 90% ($I(\mathbf{x}) < -900\,\mathrm{HU}$) the lung tissue is considered as hyperinflated, between 50% and 10%, i.e., $I(\mathbf{x}) \in [-500, -100)\,\mathrm{HU}$, as poorly aerated, and below 10% ($I(\mathbf{x}) \geq -100\,\mathrm{HU}$) as non-aerated [3]. In healthy lungs the hyperinflated interval, which is the focus of our study, is only observed in large airways, but it can also be encountered in the stomach, in the esophagus, and obviously outside the patient's body.

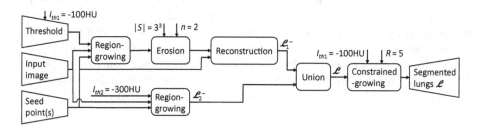

**Fig. 3.** Segmentation flowchart. Inputs in trapezes are selected by the user; the other parameters were preset: default thresholds $I_{th1} = -100\,\mathrm{HU}$ and $I_{th2} = -300\,\mathrm{HU}$, structuring-element size $|S| = 3^3$, number of iterations $n = 2$, and ball radius $R = 5$ for the constrained region-growing

In mechanically ventilated lungs, hyperinflated regions may occur at any location; therefore, our approach consists in segmenting out the aerated region ($I(\mathbf{x}) < -100\,\mathrm{HU}$) connected to the trachea (see flowchart Fig. 3), and then quantifying its sub-regions that meet the criterion $I(\mathbf{x}) < -900\,\mathrm{HU}$, regardless their connectivity. Hence, the initial segmentation is based on region-growing that starts from a seed placed by the user in the trachea, and stops when no more connected voxels below the threshold $I_{th1} = -100\,\mathrm{HU}$ can be added to the region. If visual inspection of the result detects large aerated regions disconnected from the trachea (typically by a mucus plug in a bronchus), additional seeds may be interactively selected and the segmentation may be re-run.

Unfortunately, density intervals of various tissues overlap, and fat contained within the rib-cage muscles may fall below $I_{th1}$, which results in segmentation leaks between ribs towards subcutaneous fat and even out of the body. To break these leaks, morphological erosion is performed using a structuring element of $3^3$ voxels and the number of iterations empirically set to two, and then the lung region is selected by morphological reconstruction using the seed(s) as a marker.

We have observed that, in very fat patients, leaks were too large, erosion with $n = 2$ iterations was not sufficient to break them, while increasing $n$ excessively altered the shape of the segmented region. Instead of seeking a complicated automatic solution, we implemented a simple interactive "little help": the user can use a slider to lower the threshold $I_{th1}$, and then restart the process. The segmentation result obtained to this point will be referred to as $\mathcal{L}_1^-$ regardless the threshold value used (default $I_{th1} = -100\,\mathrm{HU}$ or manually adjusted).

Erosion not only breaks leaks and "peels" the lung surface; it also removes thin regions within the lungs, namely small bronchi surrounded by non-aerated parenchyma. As long as these are connected to main airways, they can be recovered by an additional region-growing from the same seed(s) as the previous one, but using a threshold empirically set to $I_{th2} = -300\,\mathrm{HU}$, so as to penetrate far enough into the small bronchi, while avoiding leaks. The regions $\mathcal{L}_2^-$ thus segmented are then added to the previous ones: $\mathcal{L}^- = \mathcal{L}_1^- \cup \mathcal{L}_2^-$.

The last step aims at recovering the outer layer of lung voxels (with densities below $I_{th1}$) "peeled" by erosion, as well as airway voxels with densities between $-300$ and $-100\,\mathrm{HU}$ left out by the conservative stopping criterion of the second region-growing. To this purpose, a spatially-constrained region-growing operator is used. Similar to "onion-kernel" growing [6], this ball-shaped operator starts the growth from its center and agglomerates all connected voxels such that $I(\mathbf{x}) < I_{th1}$, as long as they are located within the ball. This operator is successively placed on each surface-point of $\mathcal{L}^-$, and its radius is calculated as $R = 2n + 1$.

## 2.2   Correction

The segmentation process described in the previous section may miss isolated regions. As such regions may also be located outside of the lungs, e.g., in the esophagus, fat, etc., we leave the last word to the human expert who can edit the segmentation result using smart tools specifically adapted for this task. The main goal is to aggregate missing voxels $\mathbf{x}$ that meet the criterion $I(\mathbf{x}) < I_{th1}$. To this purpose, two similar ball-shaped tools are available.

- **fill**: This tool uses the spatially-constrained region-growing operator described in the previous section, which adds voxels meeting this criterion within the ball, only if they are connected to the user-selected ball center.

- **brush**: This tool adds all the voxels meeting the criterion within the ball, regardless their connectivity.

For both tools, the user can select the radius of the ball and choose between its application in 3D or 2D. In the latter case, the tools may be applied in axial, coronal, or sagittal planes, according to the user's choice.

In some cases, it may also be useful to erase some excess voxels included in the segmentation result $\mathcal{L}$. This can be done using a ball-shaped eraser, the size of which can also be selected by the user. Obviously, to cope with possible user mistakes, we have implemented an undo/redo button.

## 2.3  Implementation

The software was implemented using the CreaTools [1] framework [1], which builds on well-known free open-source libraries (itk, vtk, wxWidgets), and significantly speeds-up the development of medical image-analysis applications. The database was built using Girder [2] web-based platform, which also is free and open source.

Figure 4 displays the GUI page corresponding to the **Validate results** function (see Fig. 2). In this example, the lung volume undergoing the cyclic hyperinflation was as large as 45% of the tidal volume (VT). The 3D segmentation results are superimposed onto the original images in axial, sagittal, and

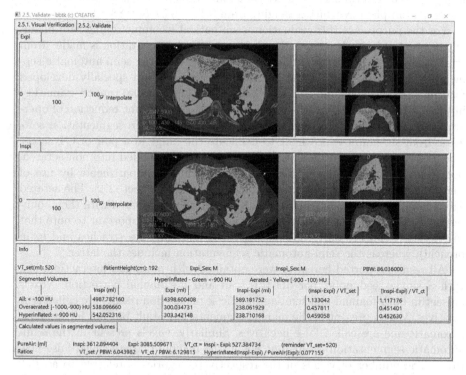

**Fig. 4.** Example of results submitted to user validation. Yellow color represents the segmentation of all aerated regions, while green color highlights the hyperinflated subregions in *expi* (upper panel) and *inspi* (central panel) images (Color figure online)

---

[1] https://www.creatis.insa-lyon.fr/site7/en/CreaTools_home.
[2] https://girder.readthedocs.io.

coronal views. The sliders can be used to adjust the opacity of the segmentation results, thus showing or hiding the underlying gray-level image. The user can explore any cross-section by moving a cross-hair, and the image contrast can be adjusted using the mouse buttons. After validation, the numerical results displayed in the lower panel are exported to a text file compatible with spreadsheets.

## 3   Experiments and Results

Our software is being evaluated within a prospective study on ARDS patients from the intensive care unit of the La Croix-Rousse Hospital, Lyon France. Hereafter, we will summarize the results obtained on data from the first twenty patients, half of them being diagnosed as Covid-19 positives. There are two 3D CT images (*inspi* and *expi*) for each patient, so the segmentation results reported correspond to forty 3D images each one containing a number of slices ranging from 238 to 384, with slice thickness of 1 mm, in-plane dimensions of $512 \times 512$ pixels, and in-plane resolution ranging from $0.57 \times 0.57$ to $0.88 \times 0.88$ mm.

For the purpose of this evaluation, each patient's complete lungs are manually segmented in both *inspi* and *expi* images. The segmentation is made "from scratch" by a medical expert blinded to the results of the semi-automatic segmentation, who uses an interactive segmentation program specially developed for this task and available from the main GUI (function `Delineate manually` in Fig. 2). Let us note that this program allowed a significant reduction of operating time, as compared to general-purpose medical-image segmentation software: from six hours to 1.5 hour to delineate the lungs in a pair of CT scans. The reference segmentation thus obtained is then subdivided into non-aerated, poorly aerated, normally aerated and hyperinflated compartments by use of the standard thresholds: $-100$, $-500$, and $-900$ HU (see Sect. 2.1). The aerated sub-regions (compartments below $-100$ HU) are then compared with their counterpart in the semi-automatic segmentation results. It is important to note that the experts delineate only the lungs, without large airways (trachea and main bronchi), whereas the semi-automatic segmentation includes the latter.

All results reported hereafter were obtained with minimum interaction, i.e., with a single seed point per 3D CT scan and no manual correction. Voxels present in both manual and semi-automatic segmentation results are called true positives. Voxels present in the manual segmentation result but not in the semi-automatic one are called false negatives. Similarly, voxels present in the semi-automatic segmentation result, but not in the manual one are called false positives. The number of voxels in the respective categories is referred to as $TP$, $FN$, and $FP$. In Table 1, these values are reported as volumes, i.e., after multiplying by voxel size. They were used to calculate performance indexes: Dice score (also known as F-measure) $Dice = 2TP/(2TP + FP + FN)$, recall (also known as sensitivity or true-positive rate) $Rec = TP/(TP + FN)$, and precision $Prec = TP/(TP + FP)$.

**Table 1.** Mean values of $TP$, $FP$, and $FN$ volumes expressed in milliliters, followed by mean values ± standard deviations of Dice score, recall, and precision. The last two columns contain Dice score and precision values calculated from $\overline{TP}$, $\overline{FP}$, and $\overline{FN}$

| Compartment | $\overline{TP}$ | $\overline{FP}$ | $\overline{FN}$ | Dice | Rec | Prec | $Dice_{tot}$ | $Prec_{tot}$ |
|---|---|---|---|---|---|---|---|---|
| Hyperinflated | 190.5 | 27.2 | 0.0 | 0.66 ± 0.27 | 1.00 ± 0.00 | 0.55 ± 0.29 | 0.94 | 0.89 |
| Normally aerated | 1969.9 | 16.8 | 2.5 | 0.99 ± 0.01 | 1.00 ± 0.01 | 0.99 ± 0.00 | 1.00 | 0.99 |
| Poorly aerated | 803.2 | 41.4 | 65.0 | 0.94 ± 0.05 | 0.94 ± 0.10 | 0.95 ± 0.03 | 0.93 | 0.95 |
| All aerated | 2963.6 | 85.4 | 67.5 | 0.97 ± 0.03 | 0.97 ± 0.06 | 0.97 ± 0.01 | 0.97 | 0.97 |

Table 2 reports the mean values of volume changes between the same patient's *inspi* and *expi* scans, calculated from manual ($M$) and semi-automatic ($S$) segmentation results. Like in the previous table, these are displayed for all compartments, as well as for the entire segmented regions (last line). These volume changes reflect the distribution of the air supplied by ventilator. In both semi-automatic and manually-obtained results, it can be seen that, while the total increase in aerated volume was – on average – below 500 ml, approximately 100 ml of this increase affected the hyperinflated compartment. Let us recall, that this corresponds to the undesired cyclic hyperinflation. Conversely, the increase in the normally-aerated volume is an expected effect, although the most desirable effect is the decrease in poorly- and non-aerated volumes, which become normally- and poorly-aerated, respectively. Unfortunately, the decrease in poorly-aerated volume was much smaller than the increase in hyperinflated volume, which may mean that VT was sometimes excessive. When comparing the semi-automatic and manually-obtained results, it can be observed that the largest difference ($D = S - M$) corresponds to the poorly-aerated compartment, which is the least contrasted with respect to the surrounding tissues, and thus the most difficult to segment. Nevertheless, although this difference was large compared to the volume change in this compartment, it remained very limited (1.2%) in comparison with the total volume of the aerated lung tissues ($L$).

**Table 2.** Mean volume changes between *inspi* and *expi* scans: $M$ stands for the results calculated from manual segmentation, $S$ for semi-automatic, $D = S - M$, and $L = TP + FN$ is the aerated lung volume from manual segmentation of the *expi* scan

| Compartment | $S$ [ml] | $M$ [ml] | $D$ [ml] | $D/M$ | $D/L$ |
|---|---|---|---|---|---|
| Hyperinflated | 101.1 | 96.0 | 5.1 | 5.3% | 0.2% |
| Normally aerated | 406.3 | 403.7 | 2.7 | 0.7% | 0.2% |
| Poorly aerated | −29.8 | −51.2 | 21.5 | −41.9% | 1.2% |
| All aerated | 477.7 | 448.5 | 29.2 | 6.5% | 1.5% |

A more precise assessment of air distribution in different compartments requires the use of Eq. 1 to calculate the quantity of air in each voxel. By summing up the contributions of each voxel, we compared the *inspi*-to-*expi* change in air quantities calculated within the segmented lungs, and the VT values set on the ventilator. The difference was equal to $-1.5\%$ on average, which is much less than the uncertainty of this setting on the ventilator (assessed by means of a calibrating device), and tends to confirm the accuracy of our segmentation.

The processing time was measured on a PC with Intel(R) Core(TM) i7-6820HQ CPU @ 2.70 GHz 8 processors and 16 GB memory. The automatic segmentation of one average-size (317 slices) 3D CT image took 46 s (except interactive selection of the seed point) and required 2.7 GB of memory. Approximately 75% of this time was spent on the constrained region-growing process. The total execution time of the application Run segmentation, i.e., loading input files for the selected patient, segmenting both *inspi* and *expi* 3D CT scans, and writing output files, was of 109 s.

## 4    Discussion and Conclusion

The focus of this study is the cyclic hyperinflation of the lungs. The proposed software was able to accurately calculate this value with user interaction limited to one click within the trachea. The mean disagreement with respect to the results obtained manually was of 5.1 ml, which represents 5.3% of the measured value. The volume of false positives in Table 1 suggests that semi-automatic segmentation tends to overestimate the aerated volume. This can be easily explained for the hyperinflated and normally-aerated compartments: whereas trachea and main bronchi were included in the semi-automatic segmentation, they were excluded from the manual one. This is not a problem in our application, where only the difference between *inspi* and *expi* volumes is of interest. Indeed, the walls of the trachea and main bronchi are quite rigid (cartilageous), so the encompassed sub-volume remains unchanged and subtraction cancels its contribution. If necessary, a strategy to remove the extra sub-volume can be implemented. Nevertheless, false positives due to the inclusion of trachea and main bronchi considerably decreased the Dice score and precision for segmentation results in the hyperinflated compartment in cases where the true volume of this compartment was close to zero. This is reflected by low mean values and large standard deviations. For this reason, we also reported the indexes $Dice_{tot}$ and $Prec_{tot}$ calculated from the mean values of $TP$, $FP$, and $FN$. These show that the overall agreement with manual segmentation was good in all compartments.

Please note that the volume of false negatives in hyperinflated and normally aerated compartments was close to zero, which means that almost no voxels were missed. Larger false-negative volumes (comparable with false positives) were observed only in the poorly-aerated compartment. As image contrasts between these regions and the extra-pulmonary tissues are considerably decreased, both semi-automatic and manual delineation of their outer boundary is more difficult. Therefore, the quality of the semi-automatic segmentation may be poorer, but the reliability of the manual reference may also be questioned.

Although validation of the proposed software is still ongoing on a larger cohort of patients, the preliminary results reported in this article are very promising, and the clinicians have begun to use it for the purpose of comparison between ARDS patients diagnosed with Covid-19 and those without the virus. The algorithms implemented in the software do not attempt to delineate the non-aerated regions, and state-of-the-art methods also fail to satisfactorily perform this task [5]. Our current investigation aims at overcoming this limitation, as quantifying the *expi*-to-*inspi* changes in this compartment would allow the assessment of another important ventilatory parameter: the alveolar recruitment.

In conclusion, so far analyzed results on 20 ARDS patients show that the proposed software meets the clinicians' expectations in terms of processing time, accuracy, and user-friendly interactive tools. For the first time, CT scans can be used to calculate the lung volume subjected to cyclic hyperinflation within a few minutes. The underlying semi-automatic segmentation can be performed by a non-expert medical operator. The next step is to integrate this tool into a real-time ventilator-adjustment strategy.

# References

1. Dávila Serrano, E.E., et al.: CreaTools: a framework to develop medical image processing software: application to simulate pipeline stent deployment in intracranial vessels with aneurysms. In: Bolc, L., Tadeusiewicz, R., Chmielewski, L.J., Wojciechowski, K. (eds.) ICCVG 2012. LNCS, vol. 7594, pp. 55–62. Springer, Heidelberg (2012). https://doi.org/10.1007/978-3-642-33564-8_7
2. Gattinoni, L., Caironi, P., Pelosi, P., Goodman, L.R.: What has computed tomography taught us about the acute respiratory distress syndrome? Am. J. Respir. Crit. Care Med. **164**(9), 1701–11 (2001). https://doi.org/10.1164/ajrccm.164.9.2103121
3. Gattinoni, L., et al.: Lung recruitment in patients with the acute respiratory distress syndrome. New Engl. J. Med. **354**(17), 1775–1786 (2006). https://doi.org/10.1056/nejmoa052052
4. Gerard, S.E., Herrmann, J., Kaczka, D.W., Musch, G., Fernandez-Bustamante, A., Reinhardt, J.M.: Multi-resolution convolutional neural networks for fully automated segmentation of acutely injured lungs in multiple species. Med. Image Anal. **60**, 101592 (2019). https://doi.org/10.1016/j.media.2019.101592
5. Klapsing, P., Herrmann, P., Quintel, M., Moerer, O.: Automatic quantitative computed tomography segmentation and analysis of aerated lung volumes in acute respiratory distress syndrome - a comparative diagnostic study. J. Crit. Care **42**, 184–191 (2017). https://doi.org/10.1016/j.jcrc.2016.11.001
6. Lidayová, K., Gómez Betancur, D.A., Frimmel, H., Hernández Hoyos, M., Orkisz, M., Smedby, Ö.: Airway-tree segmentation in subjects with acute respiratory distress syndrome. In: Sharma, P., Bianchi, F.M. (eds.) SCIA 2017. LNCS, vol. 10270, pp. 76–87. Springer, Cham (2017). https://doi.org/10.1007/978-3-319-59129-2_7
7. Mansoor, A., et al.: Segmentation and image analysis of abnormal lungs at CT: current approaches, challenges, and future trends. RadioGraphics **35**(4), 1056–1076 (2015). https://doi.org/10.1148/rg.2015140232
8. Morales Pinzón, A., Orkisz, M., Richard, J.C., Hernández Hoyos, M.: Lung segmentation by cascade registration. IRBM 38(5), 266–280 (2017). https://doi.org/10.1016/j.irbm.2017.07.003

9. Nieman, G.F., Satalin, J., Andrews, P., Aiash, H., Habashi, N.M., Gatto, L.A.: Personalizing mechanical ventilation according to physiologic parameters to stabilize alveoli and minimize ventilator induced lung injury (VILI). Intensive Care Med. Exp. 5(1), 1–21 (2017). https://doi.org/10.1186/s40635-017-0121-x
10. Yin, Y., Hoffman, E.A., Lin, C.L.: Mass preserving nonrigid registration of CT lung images using cubic B-spline. Med. Phys. 36(9), 4213–4222 (2009). https://doi.org/10.1118/1.3193526

# Edge-Aware Color Image Manipulation by Combination of Low-Pass Linear Filter and Morphological Processing of Its Residuals

Marcin Iwanowski[✉][iD]

Institute of Control and Industrial Electronics, Warsaw University of Technology,
ul. Koszykowa 75, 00-662 Warsaw, Poland
iwanowski@ee.pw.edu.pl

**Abstract.** In the paper, a method of edge-preserving processing of color images, called LPMPR (Low-Pass filter with Morphologically Processed Residuals), is proposed. It combines linear low-pass filtering with non-linear techniques, that allow for selecting meaningful regions of the image, where edges should be preserved. The selection of those regions is based on morphological processing of the linear filter residuals and aims to find meaningful regions characterized by edges of high amplitude and appropriate size. To find them, two methods of morphological image processing are used: reconstruction operator and area opening. The meaningful reconstructed regions are finally combined with the low-pass filtering result to recover the edges' original shape. Besides, the method allows for controlling the contrast of the output image. The processing result depends on four parameters, the choice of which allows for adjusting the processed image to particular requirements. Results of experiments, showing example filtering results, are also presented in the paper.

**Keywords:** Image processing · Image filtering · Mathematical morphology · Edge-preserving smoothing · Contrast enhancement

## 1 Introduction

The image low-pass filtering using linear (convolution) filters is one of the simplest, oldest, and most often applied image processing techniques. It allows for image blurring that makes image regions smooth by reducing the high-frequency image components. There are two kinds of those components: irrelevant noise and meaningful edges. To the first kind belong all image content, the blurring of which is desirable: real noise and small objects referred to, e.g. details without any importance, skin defects on portrait images, the texture of land cover elements, etc. The second kind of high-frequency component refers to rapid changes of image intensity value related to boundaries of objects that usually should be

© Springer Nature Switzerland AG 2020
L. J. Chmielewski et al. (Eds.): ICCVG 2020, LNCS 12334, pp. 59–71, 2020.
https://doi.org/10.1007/978-3-030-59006-2_6

visible without any blurring influencing contrast at edges. The classic linear low-pass filters modify both high-frequency components.

The proposed LPMPR (Low-Pass filter with Morphologically Processed Residuals) method combines the linear low-pass filtering with non-linear processing of its residual in order to find and restore the meaningful edges on the low-pass filtered image[1] and is based on the concept described in [6]. The workflow of the original approach is as follows. At first, the image is filtered using the Gaussian filter. Next, its residual is computed and split into two fractions referred to subsets of pixels of the positive and the absolute value of negative fraction. Both fractions are subject to thresholding to find strong edges (high-contrast ones). The thresholding result is next used as markers for the morphological reconstruction of residuals. Reconstructed residuals that do not contain irrelevant image components are added to the low-pass filtered image to recover meaningful edges.

In this paper, the above method is extended by adding three essential functionalities, that overcomes three imperfections of the original. The first one refers to selecting the relevant regions that have originally been based on the residual only amplitude. In the proposed approach, it is extended by adding the second, size-criterion. Secondly, the original method does not allow to control the contrast of the output image. To solve this problem, a proposed solution introduces an additional parameter – a contrast coefficient that provides for increasing/reducing the contrast of the output image. Finally, last but not least, the original method was restricted to gray-tone images only. In this paper, a way of extending it to color images is proposed.

The paper is organized as follows. Section 2 contains the discussion on related papers. In Sect. 3, the graytone case is described. Section 4 is dedicated to color image processing using the proposed approach. In Sect. 5, the examples are shown. Finally, Sect. 6 concludes the paper.

## 2   Related Work

The problem of edge-preservation in smoothing filters has already been addressed in the literature in numerous papers. The Kuwahara-Nagao filters [7,8], are based on the division of the pixel neighborhood into four overlapping regions. Based on the standard deviation, computed within these regions, the appropriate region-mean is chosen as the filter response.

The anisotropic diffusion approach [9] is based on the gradients of the filtering image to restrict the diffusion process that prevents edge blurring. This approach also exploits an idea typical for many other edge-preserving methods – it introduces a guidance image, according to which there exists a supplementary image that is used to pass the information on the way the output pixel value is computed.

---

[1] The MATLAB code of the proposed filtering method is available at https://www.mathworks.com/matlabcentral/fileexchange/77581-lpmpr-image-filter.

One of the most popular filters, based on a similar concept, is a bilateral filter [2,12]. In this filter, typical for the linear filters, the weighting has been modified. The special factor that depends on the pixel values differences has been added. It allowed for reducing the blurring in the edge regions. Among other filters of similar type are [5,10]. Relations between most popular edge-preserving filters has been studied in [1].

In [13], the Gaussian filter residuals are linearly processed in an iterative way to get better preservation of edges in the final image. An application of morphological processing to edge-aware filtering was described in [3]. The method is based on adaptive geodesic time propagation.

Among recent developments in this field, one can mention [4], where a flexible filter based on a linear model was proposed. It that may be used not only to image filtering, but to perform also some other image processing tasks like e.g. transfer structures of guidance image to the output one etc.

## 3    Graylevel Images

### 3.1    The Original Approach

The original approach [6] has been proposed for gray-level images. It is based on the residual of the Gaussian filter, denoted as:

$$I_{lf} = I_{in} * \mathcal{L}, \tag{1}$$

where $I_{in}$ stands for the original gray-level image, $*$ is the convolution operator and $\mathcal{L}$ stands for the mask of the Gaussian (or any other linear low-pass) filter. The linear filter residual is defined as:

$$I_{res} = I_{in} - I_{lf} \tag{2}$$

Further processing of the residual is based on operators that are defined on positive-valued images. Due to this fact, the $I_{res}$, which consists of both positive and negative values, is split into two fractions – positive and negative:

$$\forall p \; I_{res+}(p) = \begin{cases} I_{res}(p) & \text{if } I_{res}(p) > 0 \\ 0 & \text{if } I_{res}(p) \leq 0 \end{cases} \tag{3}$$

$$\forall p \; I_{res-}(p) = \begin{cases} -I_{res}(p) & \text{if } I_{res}(p) < 0 \\ 0 & \text{if } I_{res}(p) \geq 0 \end{cases} \tag{4}$$

Both fractions of the residual ($I_{res+}$, $I_{res-}$) are further processed in order to cut-off the irrelevant variations of the residual while preserving significant ones based on the amplitude of the residual. The processing rests on the morphological operator $\mathcal{M}$, based on the reconstruction that selects relevant parts of the residuum and preserves their original structure. Finally, relevant parts of the residuals are added to the filtered image so that meaningful edges are recovered, while the image remains blurred within the irrelevant image regions:

$$I_{out} = I_{lf} + \mathcal{M}(I_{res+}) - \mathcal{M}(I_{res-}). \tag{5}$$

The operator $\mathcal{M}$, which is the same for both residuals is defined as:

$$\mathcal{M}(I) = R_I \left( \min \left( I, \mathcal{S}_t(I) |_{\{min\{I\}, max\{I\}\}} \right) \right), \tag{6}$$

where $R_I(A)$ stands for the morphological reconstruction of the gray-level mask image $I$ from (also gray-level) markers $A$, '|' is a mapping operator that converts a binary image into gray-level one by replacing original values of 0's and 1's by two given ones. Finally, 'min' is a point-wise minimum operator of two images. The most important element of the definition of $\mathcal{M}$ (Eq. 6) is the binary marker $\mathcal{S}$ that contains meaningful, relevant regions where contrast should be preserved. The choice of these regions is based on the amplitude of the residuum:

$$\mathcal{S}_t(I) = (I \geq t). \tag{7}$$

$\mathcal{S}$ is thus the selection operator that extracts (in a form of a binary image) regions of $I$ of amplitude higher than given threshold $t$.

(a)                (b)                (c)                (d)

**Fig. 1.** Simple graylevel image filtering: original image (a), result of the Gaussian filtering with $\sigma = 15$ (b), filtering with $t = 0.1$ positive/negative mask (c) and filtering result (d)

An example of image filtering using the above approach is shown in Fig. 1. The figure exhibits results of filtering of a test image, with various $t$ values. In addition, the binary mask (markers of the meaningful regions) are drawn as white (positive binary mask: $\mathcal{S}_t(I_{res+})$) and black (negative one: $\mathcal{S}_t(I_{res-})$), gray is used to indicate regions where both masks are equal to 0. The increase of the threshold results in lower number of detected regions that are further reconstructed and used to recover the original content of the image. Finally, the number of regions where the original sharpness of regions is recovered, is decreasing.

## 3.2   Meaningful Region Selection Using Size Criterion

In the original workflow of the method, binary masks, obtained by thresholding, contain boundaries of meaningful regions. The 'meaningfulness' is estimated by means of the amplitude of residuals. However, in fact, the amplitude in not the only factor that determines the importance of image region. One may easily imagine image elements characterized by high amplitude of residual that are not important for image understanding. For example, addition of a salt-and-pepper noise to the original image introduces a lot of tiny (one-pixel-size) image elements of high amplitude on the input image and consequently modifies the residual image by adding high-amplitude elements. As such, they would be detected as meaningful regions, which is not desirable.

In order to tackle with this problem, an additional step is added to the original method. The binary mask is filtered using the area opening filter [11]. This filter removes from the binary image, all connected components of size lower that a given size threshold $s$ (size-coefficient). The equation 7 is thus extended to:

$$\mathcal{S}_{t,s}(I) = (I \geq t) \circ (s),\tag{8}$$

where $\circ(s)$ is the area opening, removing components smaller than given size defined by $s$ (number of pixels in the connected component of the thresholded fraction of the residual).

$\quad$ (a) $\qquad\qquad$ (b) $\qquad\qquad$ (c) $\qquad\qquad$ (d)

**Fig. 2.** Results of filtering for various $s$ values – Gaussian with $\sigma = 15$, $t = 0.05$: positive/negative mask and filtering result for $s = 20$ (a), (b); $s = 110$ (c), (d)

$\quad$ The result of applying the size coefficient $s$ is shown in Fig. 2. Comparing to Fig. 1 one may observe that in both cases, when the coefficient ($t$ in case of Fig. 1 and $s$ in case of Fig. 2) increases, the number of selected regions decreases. Contrary, however, to the original case, when coefficient $s$ is used, the regions are removed based on their size. It allows thus to reject small – in terms of a number of pixels – objects from the residual even if their amplitude is high. Finally, it allows for keeping these regions blurred on the final image.

### 3.3   Contrast Control

One of the classic approaches to image contrast enhancement is the based on the subtraction (or addition – depending on the filter mask coefficients) of the high-pass filter from the image itself. It is based on the high-pass filter property that allows for detecting the local variations of image pixel values. The difference between low-pass filter and the image itself is another way of getting the high-pass filtering result. In the proposed method, the morphologically-processed residuals refer to image high-frequency components containing regions of the amplitude above the threshold $t$. In order to get control of the output image contrast, a *contrast control* coefficient $c$ is introduced in the Eq. 5, so that the modified formula is defined now as follows:

$$I_{out} = I_{lf} + (\mathcal{M}(I_{res+}) - \mathcal{M}(I_{res-}))) \cdot c. \tag{9}$$

Depending on $c$, the contrast is either enhanced ($c > 1$), preserved ($c = 1$) or reduced ($0 < c < 1$). An example showing how the contrast control coefficient affects the result of processing is shown in Fig. 3. In addition, the contrast enhancement result *without* morphological processing of residuals is shown. Comparing picture Fig. 3(c) and (d) one may see how the morphological processing affects on the number of small image details visible on the final image. Proposed processing allows for rejecting details of low importance so that finally only contrast of meaningful regions is enhanced.

(a)                    (b)                    (c)                    (d)

**Fig. 3.** The influence of the contrast factor $c$ to the final result. The original 'Lena' image (a), LPMPR filtering without contrast modification, $c = 1, t = 0.2$ (b), contrast enhancement $c = 1.5, t = 0.2$ (c), contrast enhancement $c = 1.5$ without residual filtering ($t = 0$) (d)

## 4   Color Image Processing

The proposed method is based on both linear and non-linear techniques. The extension of linear methods from gray-level case to color one is usually achieved by independent processing of color components. Contrary to the linear case, the ways of extending non-linear techniques to color case are more differentiated.

The main reason are problems caused by non-linear independent processing of separated channels. They usually appear as modification of colors within the final color image, that in worst case have a form of so-called false colors i.e. colors that are not present on the original image.

To solve this problem, an application of a *single* binary mask, the same for all channel is proposed. Thanks to that, for each color component, the same regions are marked as meaningful, and consequently the same regions are reconstructed on all color channels. The proposed way of producing a single mask is based on the assumption that the primary role for interpreting the presence of objects on the image plays *image luminance*. For the input RGB color space, it is computed as the weighted average of components. Thanks to the linear properties of the first part of processing, instead of starting from the luminance of the input image, the luminance of residuals is computed giving the same result:

$$L_{res+} = lum(I_{res+}); \; L_{res-} = lum(I_{res-}), \tag{10}$$

where upper indexes R, G, B determines the color component and $lum(I) = 0.3I^R + 0.6I^G + 0.1I^B$ is the luminance.

(a)                    (b)                    (c)                    (d)

**Fig. 4.** Color processing (Gaussian with $\sigma = 11$, $t = 0.12$, $c = 1.8$, $s = 0$). Original image (a), blurred image (b), independent processing of color channels (c), luminance-driven processing (d) (Color figure online)

An example of color processing is shown in Fig. 4 – the quality of image obtained by independent channel processing is visibly lower than the result of applying the proposed approach.

Considering all described extensions, the complete processing scheme of color images may be summarized by following algorithm:

- Input color image $I_{in}$, output color image $I_{out}$
- Parameters: linear filter mask $\mathcal{L}$, amplitude threshold $t$, contrast coefficient $c$, size threshold $s$

1. Filter $I_{in}$ using linear filter with mask $\mathcal{L}$ to get $I_{lf}$ (Eq. 1)
2. Compute $I_{res}$, $I_{res+}$, $I_{res-}$ (Eq. 2, 3, 4)
3. Compute residuals of the luminance $L_{res+}$ and $L_{res-}$ (Eq. 2)
4. For each residual ($L_{res+}$, $L_{res-}$) create binary mask ($\mathcal{S}_{t,s}(L_{res+})$, $\mathcal{S}_{t,s}(L_{res-})$) applying given $t$ and $s$ (Eq. 8)
5. For each $cc \in \{R, G, B\}$ perform reconstruction of $I_{res-}^{cc}$ with mask $\mathcal{S}_{t,s}(L_{res-})$ and get $\mathcal{M}(I_{res-}^{cc})$ (Eq. 6)
6. Do the same using $I_{res+}^{cc}$ and $\mathcal{S}_{t,s}(L_{res+})$ to get $\mathcal{M}(I_{res+}^{cc})$ (Eq. 6)
7. Combine residuals with $I_{filt}$ and get final image $I_{out}$ (Eq. 9) for given $c$

The control of the proposed method is performed by means of three parameters $t$, $s$ and $c$. Depending on their choice one controls level of details recovered based on the amplitude of residual ($t$), as well as the size of particles residual ($s$). Moreover, thanks to parameter $c$ one controls also the contrast of the output image. An example showing how the choice of parameters $t$ and $s$ influences the final result is shown in Fig. 5.

(a)                    (b)                    (c)                    (d)

**Fig. 5.** Results of processing with various parameter sets ($c = 1.5$ and $\sigma = 5$ are constant): $t = 0.05$, $s = 0$ (a); $t = 0.15$, $s = 0$ (b); $t = 0.15$, $s = 2$ (c); $t = 0.15$, $s = 8$ (d)

## 5   Results

The comparison of the proposed method with guided and bilateral filters (with default parameters in their MATLAB implementations) is shown in Fig. 6. The proposed approach allows for better extracting the relevant object (not blurred) from the blurred background. In the case of its competitors the increase of the expected blurring would affect also to quality and sharpness of relevant object.

**Fig. 6.** Original image (a), filtered using the proposed method (b), using guided filter (c) and bilateral filter (d) (Color figure online)

Due to its four parameters, the LPMPR filters are flexible. They may be adjusted to get filtering results focused on preserving edges of a particular object or content classes of the specific image. The image shown in Fig. 7 is a part of a high-resolution satellite image showing a mountain house located close to the lake with different types of landcover classes around: trees, bushes, water, stones, underwater stones, grass, road, etc. Depending on the choice of parameters, various classes might remain sharp, while the remainder of the image is blurred. The original image shown at the position (a) has been filtered using multiple sets of parameters, that are printed below figures at particular positions. All the operations were performed on the original image (a) with Gaussian filtering with $\sigma = 15$ (b). The results of LPMPR are the following. Having a look at the position (c) one may see that only the smallest stones at the lakeshore are blurred (low value of $t$), in case of (d), for $t = 0.1$ similar results are obtained for lower $s$ but now, also the underwater stones disappear. On position (e) it is visible that, when increasing $s$, the area covered by bushes and grass became blurred, on (f) further increase of $t$ results in more robust filtering results. The remainder of cases illustrates the contrast issues, on (g) high values of $t$ and $s$ produces an image where most of the area is blurred apart from the house and some other high-contrasted details. The next one, (h), shows what happens when going down with $s$ and, in the same, increasing $c$. It results in a highly contrasted final image but with small details of texture blurred. The image at the last position (i) is presented to compare with (h) – it depicts the same level of contrast enhancement but without morphological processing of residuals is the result of classic contrast enhancement method.

The method may be also successively applied, to filter portrait images of human faces, where skin smoothing is one of the most popular tasks. Its application allows for reducing considerably the noise level inside the skin regions (what makes the skin smooth) and preserving the contrast on image edges (face and its details outlines). The example of such an application is shown in Fig. 8.

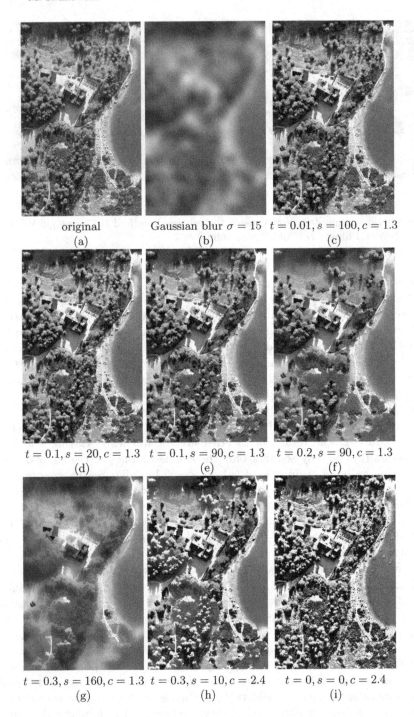

original          Gaussian blur $\sigma = 15$    $t = 0.01, s = 100, c = 1.3$
(a)                    (b)                        (c)

$t = 0.1, s = 20, c = 1.3$   $t = 0.1, s = 90, c = 1.3$   $t = 0.2, s = 90, c = 1.3$
(d)                          (e)                          (f)

$t = 0.3, s = 160, c = 1.3$   $t = 0.3, s = 10, c = 2.4$   $t = 0, s = 0, c = 2.4$
(g)                           (h)                          (i)

**Fig. 7.** LPMPR filtering results of the satellite image with various land-cover classes, for filtering parameters of see captions below images for description see the text (Color figure online)

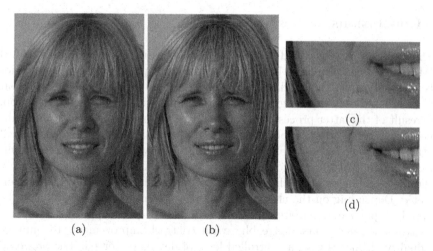

**Fig. 8.** Photo enhancement. The original picture (a), filtered one (b), enlarged part of the original (c) and of the filtered one (d)

In most of the previous examples, the contrast coefficient was set to values slightly above 0 ($1 < c < 1.8$). This results in increasing pixel values difference on edges and improve the contrast of meaningful image objects. This range of possible changes of $c$ gives, as a result, natural-looking images with visibly increased contrast of details. Further increase of $c$ ($c > 2$) causes a somehow unnatural outlook of images. The boundaries between image regions become so sharp that its color becomes either black or white. The extremely high contrast makes image object looking comics-like – flat regions of relatively uniform color value separated by black or white lines. An effect of transferring an image into comics is shown in Fig. 9.

**Fig. 9.** Two examples of comics effect. Filtering with (in both cases) with $t = 0.03$, $c = 3$, $s = 1000$.

# 6   Conclusions

In the paper, a novel method for edge-aware processing of color images, called LPMPR, has been proposed. It is based on the principle of extracting the residuals of the low-pass linear filter that is further processed using a non-linear approach based on morphological operators of area opening and reconstruction. The result of the latter processing is added back to the low-pass filtering result. Non-linear filtering aims to localize and extract meaningful edges reconstructed on the final image so that blurring is not present within their regions. Four parameters control the method's behavior: the mask of the Gaussian (on any other low-pass linear) filter, the threshold $t$, size coefficient $s$, and contrast coefficient $c$. Depending on the choice of these parameters, the final result may be adapted to particular requirements. The method may be applied to perform a controllable edge-aware image blur with contrast improvement, to improve the quality of images with a controlled level of details preserved. The proposed LPMPR method has been illustrated in the paper by examples demonstrating its usefulness.

# References

1. Barash, D.: Fundamental relationship between bilateral filtering, adaptive smoothing, and the nonlinear diffusion equation. IEEE Trans. Pattern Anal. Mach. Intell. **24**(6), 844–847 (2002)
2. Durand, F., Dorsey, J.: Fast bilateral filtering for the display of high-dynamic-range images. ACM Trans. Graph. **21**(3), 257–266 (2002)
3. Grazzini, J., Soille, P.: Edge-preserving smoothing using a similarity measure in adaptive geodesic neighbourhoods. Pattern Recogn. **42**(10), 2306–2316 (2009). Selected papers from the 14th IAPR International Conference on Discrete Geometry for Computer Imagery 2008
4. He, K., Sun, J., Tang, X.: Guided image filtering. IEEE Trans. Pattern Anal. Mach. Intell. **35**(6), 1397–1409 (2013)
5. Hong, V., Palus, H., Paulus, D.: Edge preserving filters on color images. In: Bubak, M., van Albada, G.D., Sloot, P.M.A., Dongarra, J. (eds.) ICCS 2004. LNCS, vol. 3039, pp. 34–40. Springer, Heidelberg (2004). https://doi.org/10.1007/978-3-540-25944-2_5
6. Iwanowski, M.: Morphological processing of Gaussian residuals for edge-preserving smoothing. In: Angulo, J., Velasco-Forero, S., Meyer, F. (eds.) ISMM 2017. LNCS, vol. 10225, pp. 331–341. Springer, Cham (2017). https://doi.org/10.1007/978-3-319-57240-6_27
7. Kuwahara, M., Hachimura, K., Eiho, S., Kinoshita, M.: Processing of RI-angiocardiographic images. In: Preston, K., Onoe, M. (eds.) Digital Processing of Biomedical Images, pp. 187–202. Springer, Boston (1976). https://doi.org/10.1007/978-1-4684-0769-3_13
8. Nagao, M., Matsuyama, T.: Edge preserving smoothing. Comput. Graph. Image Process. **9**(4), 394–407 (1979)
9. Perona, P., Malik, J.: Scale-space and edge detection using anisotropic diffusion. IEEE Trans. Pattern Anal. Mach. Intell. **12**(7), 629–639 (1990)

10. Saint-Marc, P., Chen, J., Medioni, G.: Adaptive smoothing: a general tool for early vision. IEEE Trans. Pattern Anal. Mach. Intell. **13**(6), 514–529 (1991)
11. Soille, P.: Morphological Image Analysis: Principles and Applications. Springer, Berlin (2004). https://doi.org/10.1007/978-3-662-05088-0
12. Tomasi, C., Manduchi, R.: Bilateral filtering for gray and color images. In: Sixth International Conference on Computer Vision (IEEE Cat. No.98CH36271), pp. 839–846, January 1998
13. Wheeler, M.D., Ikeuchi, K.: Iterative smoothed residuals: a low-pass filter for smoothing with controlled shrinkage. IEEE Trans. Pattern Anal. Mach. Intell. **18**(3), 334–337 (1996)

# On the Effect of DCE MRI Slice Thickness and Noise on Estimated Pharmacokinetic Biomarkers – A Simulation Study

Jakub Jurek[1]([✉]) [iD], Lars Reisæter[2] [iD], Marek Kociński[1,3] [iD],
and Andrzej Materka[1] [iD]

[1] Institute of Electronics, Łódź University of Technology, Łódź, Poland
jakubjurekmail@gmail.com
[2] Department of Radiology, Haukeland University Hospital, Bergen, Norway
[3] Mohn Medical Image and Visualisation Center, Department of Biomedicine,
University of Bergen, Bergen, Norway

**Abstract.** Simulation of a dynamic contrast-enhanced magnetic resonance imaging (DCE MRI) multiple sclerosis brain dataset is described. The simulated images in the implemented version have $1 \times 1 \times 1 \, \text{mm}^3$ voxel resolution and arbitrary temporal resolution. Addition of noise and simulation of thick-slice imaging is also possible. Contrast agent (Gd-DTPA) passage through tissues is modelled using the extended Tofts-Kety model. Image intensities are calculated using signal equations of the spoiled gradient echo sequence that is typically used for DCE imaging. We then use the simulated DCE images to study the impact of slice thickness and noise on the estimation of both semi- and fully-quantitative pharmacokinetic features. We show that high spatial resolution images allow significantly more accurate modelling than interpolated low resolution DCE images.

**Keywords:** DCE imaging · Quantitative DCE analysis ·
Semi-quantitative DCE analysis · Biomarkers

## 1 Introduction

Dynamic contrast-enhanced magnetic resonance imaging (DCE-MRI) is a type of perfusion imaging. DCE can be used to assess tissue microcirculation parameters that are biomarkers for diagnosis, prognosis and treatment monitoring [6]. A DCE dataset is a time sequence of T1-weighted (T1W) MR images, acquired before, during and after administration of a paramagnetic contrast agent (CA).

DCE images are characterized, among others, by spatial and temporal resolution. The spatial resolution is defined as the ability to differentiate small structures in the image, while the temporal resolution refers to the time lapse between consecutive T1W scans. There exists a trade-off between the spatial

© Springer Nature Switzerland AG 2020
L. J. Chmielewski et al. (Eds.): ICCVG 2020, LNCS 12334, pp. 72–86, 2020.
https://doi.org/10.1007/978-3-030-59006-2_7

and temporal resolution of DCE images. In general, as explained in [27], the decision concerning the acquisition of a high spatial resolution or a high temporal resolution dataset depends on the type of image analysis that is required (a review of those can be found in [15]), which in turn depends on the disease or condition. The temporal resolution would be more important than spatial if the CA kinetics are to be finely studied. If tumour heterogeneity is to be studied, spatial resolution is more important than temporal. High spatial resolution of breast cancer MRI was shown to be critical by [9].

We claim that the trade-off between spatial and temporal resolution in DCE MRI can be improved, using super-resolution methodology, for example. The proof of the above claim requires demonstration of the feasibility of super-resolution methods, preferably using quantitative evaluation. In [14], we used MRI images of a physical phantom with known geometrical properties to quantitatively evaluate super-resolution images. For evaluation of super-resolution DCE images we postulate to use a digital phantom with known pharmacokinetics to simulate high spatial resolution DCE images.

According to our literature research, there have been just a few attempts in the past to simulate realistic DCE MRI data. Pannetier et al. [21] simulated a DCE experiment on the level of cells and microvessels. Abdominal DCE MRI simulation was studied in [2], using anatomy masks derived from CT images. Similarly, [7] simulated CA-based enhancement in a DCE sequence of the abdomen acquired without CA administration, using a mathematical model. In [3], prostate cancer DCE was simulated using an anatomical atlas of the prostate. These approaches used thick-slice images of the abdomen or prostate and thus resulted in low resolution (LR) simulated DCE images that are not suitable for evaluation of super-resolution DCE images. Simulation of higher resolution images was proposed by [4], which used a digital phantom of the brain, but did not model the partial volume effect (PVE), responsible for blurring and loss of spatial resolution. In this paper, we simulate DCE images by using the DCE signal modelling methods similar to [2,3,7] with a high resolution (HR) brain phantom published on BrainWeb: Simulated Brain Database [5], similarly as [4]. The BrainWeb phantom, however, has an important property that was not exploited in [4]: apart from having relatively high, isotropic spatial resolution, it is a fuzzy model with knowledge about the partial volume effect within each voxel. We innovate DCE image simulation by modelling the impact of PVE on enhancement curves in the high resolution isotropic simulated images and in thick-slice LR simulated images. We then conduct a study to estimate quantitative and semi-quantitative biomarkers from these images. We focus on the impact of slice thickness and noise on the value of the estimates.

## 2 Materials

### 2.1 Anatomical Phantom of the Brain

Construction of the adapted BrainWeb digital brain phantom is described in [5]. The phantom dataset is composed of 11 tissue maps of voxel resolution

$1 \times 1 \times 1\,\mathrm{mm}^3$. The value of voxels is in the $[0, 1]$ range and denotes the proportional contribution of a particular tissue to the voxel tissue content. The tissue classes and their parameters are summarized in Table 1. A sagittal slice through the phantom image is shown in Fig. 1, where each voxel has a label referring to the tissue with largest contribution[1].

**Table 1.** Tissue parameters of the brain phantom used for MRI simulation in [17]

| Parameter/tissue class | Background | Cerebrospinal fluid | Gray matter | white matter | Glial matter | Fat | Muscle + skin | Skin | Skull | Meat | MS lesions |
|---|---|---|---|---|---|---|---|---|---|---|---|
| T1 | 0 | 2569 | 833 | 500 | 833 | 350 | 900 | 2569 | 0 | 500 | 752 |
| T2* | 0 | 58 | 69 | 61 | 69 | 58 | 30 | 58 | 0 | 61 | 204 |
| PD | 0 | 1 | 0.86 | 0.77 | 0.86 | 1 | 1 | 1 | 0 | 0.77 | 0.76 |

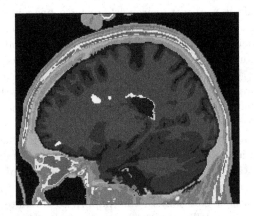

**Fig. 1.** The BrainWeb phantom shown as a tissue-labelled sagittal slice

## 3   Methods

### 3.1   Modelling of CA Concentration in Blood Plasma

The organs are supplied in blood by at least one artery. The model of the curve showing CA blood plasma concentration over time is called the arterial input function (AIF). Personalized models assume that the parameters of the AIF curve can be found by fitting the model to DCE timeseries taken from a large artery. Other models include population-based models, that are obtained by averaging of CA blood plasma contrast measurements of a group of subjects. A common population-based model is the Tofts biexponential model [12,25],

---

[1] https://brainweb.bic.mni.mcgill.ca/tissue_mr_parameters.txt.

used for DCE simulation by [3]. The temporal resolution of the data that were used to fit the Tofts model, however, was several minutes [26]. This timescale does not match the temporal resolution of modern DCE images. We therefore use the Parker AIF [22], given by

$$C_b(t) = \sum_{n=1}^{2} \frac{A_n}{\sigma_n \sqrt{2\pi}} \exp(-(t-T_n)^2/2\sigma_n^2) + \alpha \exp(-\beta t)/(1+\exp(-s(t-\tau))) \quad (1)$$

where the CA concentration in blood $C_b(t)$ is a parametrised sum of two Gaussians and an exponential modulated by a sigmoid function. The parameters were found in [22] by fitting the model to several DCE datasets and averaging: $A_1 = 0.809$ mMol · min, $A_2 = 0.330$ mMol · min, $T_1 = 0.17046$ min, $T_2 = 0.365$ min, $\sigma_1 = 0.0563$ min, $\sigma_2 = 0.132$ min, $\alpha = 1.05$ mMol, $\beta = 0.1685$ mMol, $s = 38.078$ min$^{-1}$, $\tau = 0.483$ min$^{-1}$. The AIF is shown in Fig. 2.

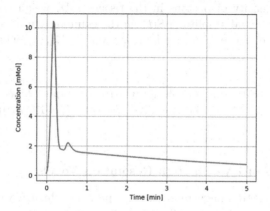

**Fig. 2.** The AIF assumed in our simulation, as proposed by Parker et al. [22]. Note that concentration peaks (first-pass- and recirculation-related) occur during the first minute of CA passage

### 3.2  Modelling of CA Concentration in Tissues

The concentration of the CA in tissues is described by the extended Tofts-Kety model:

$$C_p(t) = C_b(t)/(1-H) \quad (2)$$

$$C_t(t) = v_p C_p(t) + C_p(t) \otimes (K^{trans} e^{-k_{ep}t}) \quad (3)$$

Equation (2) calculates blood plasma CA concentration $C_p(t)$ using the blood hematocrit value $H = 0.42$ and CA concentration in blood $C_b(t)$. Then, Eq. (3) calculates the tissue CA concentration $C_t(t)$ which is a sum of two components dependent on the plasma concentration, where $\otimes$ denotes convolution, $K^{trans}$ is the volume transfer constant, $k_{ep}$ is the efflux rate constant from extravascular extracellular space to plasma and $v_p$ is the blood plasma volume per unit volume of tissue, respectively [19,24]. The latter three pharmacokinetic parameters are used to describe the state of the tissue and can reflect malignancy, blood-brain barrier (BBB) disruption etc.

### 3.3 From CA Concentration to MR Signal Enhancement

In general, the MR signal in T1W MR images is related to both the T1 and T2* relaxation times of tissues. The CA creates chemical bonds with signal-generating molecules and changes the native T1 and native T2 times ($T_{10}$ and $T_{20}^*$) by a factor related to the longitudinal and transverse relaxivities of the CA ($R_1$ and $R_2$) and the CA concentration in tissue. The contrast of the MR images is further dependent on the acquisition parameters set on the MR scanner: the echo time (TE), repetition time (TR) and flip angle (FA). Finally, scanner gain $k$ and the density of protons in the tissue ($PD$) weigh the signal ampli-tude. Altogether, signal amplitude $S(t)$, in the commonly used spoiled gradient echo acquisition sequence that we adapt, is given by Eq. (4), which we modified from [3,27].

$$S(t) = k \cdot PD \cdot sin(FA) \cdot exp(-TE(\tfrac{1}{T_{20}^*} + R_2 C_t(t))) \cdot (1 - exp(-TR(\tfrac{1}{T_{10}} + R_1 C_t(t)))) \\ \cdot (1 - cos(FA) \cdot exp(-TR(\tfrac{1}{T_{10}} + R_1 C_t(t))))^{-1} \tag{4}$$

### 3.4 Modelling of PVE in High Resolution Voxels

Each voxel of an MR image represents a space in which a variety of tissues is present. These tissues have different $PD$ value and relaxation times. For a particular voxel, its intensity in the image depends on the sum of signals sent by the mixture of tissues. Knowing the proportion of the tissues contributing to each voxel and the $PD$, we propose to calculate this sum using Eq. (5):

$$S^v(t) = \sum_{l=1}^{11} \lambda_l^v \cdot k \cdot PD_l \cdot a_l(t) \tag{5}$$

where $v$ is the index of the current voxel, $l$ is the tissue class label and $\lambda_l^v$ is the contribution of tissue $l$ to voxel $v$. Using the above formula, PVE can be modelled for the original $1 \times 1 \times 1\,mm^3$-sized voxels of the HR phantom.

### 3.5 Modelling of Thick-Slice Imaging

To model a low spatial resolution DCE imaging of the brain, we used an imaging model described in [14]:

$$I^{LR} = \mathbf{D}(I^{HR}) \tag{6}$$

where $\mathbf{D}$ is an operator leading to $I^{LR}$, a LR thick-slice anisotropic-voxel image and $I^{HR}$ is the high resolution isotropic volume.

Operator $\mathbf{D}$ takes as input a block of $1 \times 1 \times AF$ neighbouring isotropic voxels (in the single, desired direction), calculates their average intensity and assigns it to the anisotropic output voxel, which covers the same space as the input voxel block. One can also view these operations as averaging $AF$ thin slices to compute a single thick-slice. We refer to $AF$ as the anisotropy factor.

Downsampling and averaging leads to an increased PVE and aliasing artifacts, deteriorating image quality [13]. As we will show, it also introduces errors to the estimates of pharmacokinetic parameters.

### 3.6 Modelling of Noise

In DCE analysis, noise spoils not only the three spatial dimensions of the image, but also the fourth temporal dimension. In MRI generally, the noise has Rician distribution in high-intensity regions and Rayleigh distribution (a special case of the Rician) in background regions with no MR signal [10]. This situation can be modelled in the following way. Our simulated DCE images are noise-free, real-valued magnitude images, as follows from Eq. (5). The intensity $S_n^v$ of a noisy image voxel can be modelled as:

$$S_n^v = \sqrt{(S^v + N_R)^2 + (jN_I)^2} \qquad (7)$$

where $j$ is the imaginary unit, $N_R$ is the additive noise of the real part of the voxel signal and $N_I$ is the additive noise of the complex part of the signal. This refers to the physical mechanism of image acquisition, as explained in [16], but the model can be used for our simulation as well. Both $N_R$ and $N_I$ are random, zero-mean Gaussian noise variables. Calculating the magnitude of the complex intensity in Eq. (7) gives images with the desired Rician noise distribution.

The level of noise is defined by the standard deviation of the Gaussians $N_R$ and $N_I$, that is specified as a certain percentage of the intensity of the brightest tissue. In our experiment, we use 3% noise relative to the brightest tissue, which we find reasonable based on a comparison with a real brain DCE dataset.

### 3.7 Semi-quantitative DCE Analysis

Semi-quantitative analysis of DCE timeseries is based on the paradigm of curve shapes [8,11], that are believed to correspond to the presence and aggressiveness of neoplasms. This type of analysis does not require more than the DCE dataset itself to derive meaningful features. The features, as shown in Fig. 3, are estimated from interpolated, normalized time-intensity curves and include time-to-peak, wash-in gradient and wash-out gradient, among others [11]. Time-to-peak is defined as the time necessary for the voxel signal to reach its maximum value $(t_3, S_{100\%})$. Wash-in gradient is the rate of change of the intensity starting at the timepoint where the intensity has 10% of the maximal value $(t_1, S_{10\%})$,

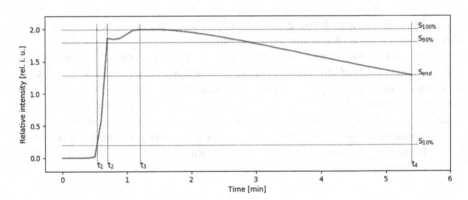

**Fig. 3.** Relevant times and relative signal intensity values for the determination of semi-quantitative pharmacokinetic parameters [11]

until it reaches 90% of the maximum value $(t_2, S_{90\%})$

$$G_{in} = \frac{t_2 - t_1}{S_{90\%} - S_{10\%}} \tag{8}$$

Wash-out gradient is defined as the rate of change of the intensity after time-to-peak until the end of the timeseries $(t_4, S_{end})$.

$$G_{out} = \frac{t_4 - t_3}{S_{end} - S_{100\%}} \tag{9}$$

Normalization of the curves is achieved by subtracting the mean value of the pre-injection signal from all time points and then dividing by the same mean value [11]. The injection time can be determined manually or automatically, while in our experiments it is known by design. For interpolation of the 6-s resolution simulated timeseries we used linear interpolation and subsampling by a factor of 60, thus obtaining the temporal resolution of 1/10 s for semi-quantitative analysis.

Semi-quantitative analysis seems to be more robust than fully quantitative analysis as demonstrated in [18]. We believe this is mainly due to the number of unknown variables in Eq. 4. We shall leave investigation of this problem to another study.

## 3.8 Fully Quantitative DCE Analysis

Fully quantitative analysis is a way to estimate the value of $K^{trans}$, $k_{ep}$ and $v_p$ parameters by fitting the function from Eq. (3) to the measured time-intensity timeseries converted to time-concentration curves, for every voxel. To obtain the time-concentration curves, it is not enough to acquire the DCE dataset. Conversion requires the estimation of the AIF for the analysed patient, which can be done in two ways. The first way is to use an averaged, population-based input function, such as the ones proposed by Orton [20] or Tofts-Weinmann [25,26]. The advantage is that they are readily available, in contrast to personalized AIFs. To measure the latter, it is required to capture a large vessel in the image field-of-view. The size of the vessel is important due to smaller expected PVE in larger vessels, which leads to better AIF parameter estimation.

If the AIF is known, the remaining variables necessary to obtain the time-concentration curves are $R_1$, $R_2$, TE, TR, FA (known by design) and $k$, $PD$, $T_{20}^*$ and $T_{10}$ that still need to be estimated. The relaxation times can be estimated using multiple variable flip angle (VFA) acquisitions and appropriate signal equations [16]. However, it would be infeasible to extend scanning time in practice to estimate both $T_{20}^*$ and $T_{10}$. The first of these appears in the term $exp(-TE(\frac{1}{T_{20}^*} + R_2C_t(t)))$ in (4). We obtain

$$exp(-TE(\frac{1}{T_{20}^*} + 0)) = exp(\frac{-TE}{T_{20}^*}) \qquad (10)$$

for $t = 0$ and

$$exp(-TE(\frac{1}{T_{20}^*} + R_2C_t(t))) = exp(\frac{-TE}{T_{20}^*}) \cdot exp(-TE \cdot R_2C_t(t)) \qquad (11)$$

for $t > 0$. The influence of (10) component on the total signal (4) can be neglected if TE is set to be short, given that $T_{20}^*$ of tissues is longer than 30 ms [16,27] (Table 1). Then, $T_{10}$ is estimated by acquiring at least two T1-weighted datasets with a different FA, keeping the TR and TE constant. [27] suggests that $k \cdot PD$ and $T_{10}$ can be fit to the signal equation, while we calculate it directly using only two VFA acquisitions. Then,

$$\frac{T_{10}}{TR} = log_e(\frac{S_1 sin(FA_2)cos(FA_1) - S_2 sin(FA_1)cos(FA_2)}{S_1 sin(FA_2) - S_2 sin(FA_1)})^{-1} \qquad (12)$$

where $S_1$ and $S_2$ are intensities of the two T1-weighted images and $FA_1$, $FA_2$ are two different flip angles. Once $T_{10}$ is estimated, it can be used to calculate $k \cdot PD$ by solving Eq. (4) for $t = 0$ and $TE \gg T_{20}^*$. With all the other parameters in hand, $K^{trans}$, $k_{ep}$ and $v_e$ can finally be estimated.

The latter yet requires conversion of the measured time-intensity signals to time-concentration signals. The relation is obtained from (4) and is the following:

$$C(t) = (TR + log_e(1 - kPDsin(FA)S^{-1}) - log_e(cos(FA) \\ - kPDsin(FA)S^{-1})) \cdot (-R_1T_{10}TR)^{-1} \qquad (13)$$

We use a curve fitting approach to find the parameters of (3). The Levenberg-Marquardt algorithm was used to minimize the least squares error between the converted observed data and the model (3).

### 3.9 Implementation

All models were implemented in-house using Python and its popular libraries such as NumPy, SciPy, NiBabel, scikit-image and Matplotlib. Quantitative and semi-quantitative parameters were computed likewise. The programs were run on a laptop with an Intel Core i3-4000M CPU with 2.4 GHz clock speed and 16 GB of random access memory.

### 3.10  Simultion of High Spatial Resolution VFA and DCE Images

Our simulated HR scanning session included acquisition of 56 image volumes in total. For most parameters of the acquisition, we followed the Quantitative Imaging Biomarkers Alliance 1.0 profile for DCE quantification [1]. The first VFA $T_1$-weighted image was simulated using $FA = 15°$, $TR/TE = 4\,ms/1\,ms$, scanner gain $k$ was set to 10000 for all simulations. The DCE sequence was simulated using (5) for every voxel of the phantom, with TR/TE unchanged and $FA = 25°$.

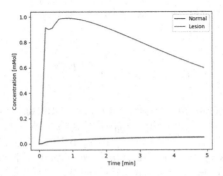

**Fig. 4.** Model-based pharmacokinetic curves representing CA concentration changes over time in normal (blue) and lesion (red) white matter (WM) tissue (Color figure online)

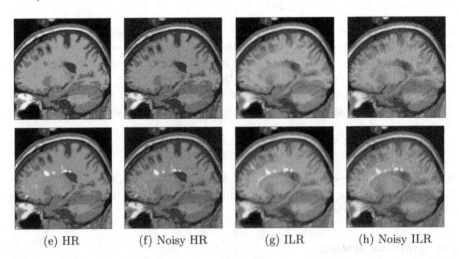

(e) HR          (f) Noisy HR          (g) ILR          (h) Noisy ILR

**Fig. 5.** Sample time-frames (pre-contrast - top, post-contrast - bottom) from simulated HR and interpolated LR (ILR) DCE images (Color figure online)

The first time-frame of the DCE sequence was used as the second VFA image for $T_1$-mapping, using (12). In total, the DCE sequence consisted of 55 volumes, 5 of which were pre-contrast ones. The simulated CA was gadopentate dimeglumine (Gd-DTPA) with $R_1 = 3.75\,(\mathrm{mM \cdot s})^{-1}$ and $R_2 = 4.89\,(\mathrm{mM \cdot s})^{-1}$[23]. The temporal resolution was set to 6 s, although the convolution in (3) was performed with signals of 1 s temporal resolution and the result was downsampled. For all tissues but the MS lesions, we set $K^{trans} = 0.0002$, $k_{ep} = 0.2$, $v_p = 0$. For MS lesions, which we assumed to be fully homogeneous, we set $K^{trans} = 0.01$, $k_{ep} = 1$, $v_p = 0.03$. The resulting ground-truth concentration curves for lesions and normal tissue are shown in Fig. 4. Pre-contrast and post-contrast slices of the resulting DCE volumes are shown in Fig. 5.

### 3.11  Simulation of Low Spatial Resolution DCE Images

Low spatial resolution VFA volumes and DCE sequence were obtained from the HR counterparts using (6) with $AF = 6$. 3% noise was added to the volumes following LR imaging simulation. The LR images were then linearly interpolated to recover the original voxel resolution, although with a loss in the spatial resolution. As demonstrated in Fig. 5, the ILR images are blurred in the plane of the slices and aliasing might occur in the direction perpendicular to the slices.

## 4  Results

### 4.1  Comparison of Time-Intensity Signal Curves in HR and LR DCE Images

Curves sampled from a lesion region and from a healthy region are plotted in Fig. 6. In the case of the healthy white matter, due to the protection of the BBB, only little enhancement is present. Since the sample was taken from a specific

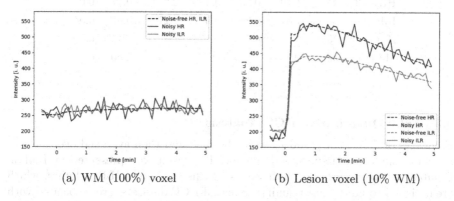

(a) WM (100%) voxel          (b) Lesion voxel (10% WM)

**Fig. 6.** Sample intensity timeseries taken from noisy and noise-free HR and ILR images. In a), PVE does not occur due to tissue homogeneity and the difference in noisy curves is due to noise only. In b), PVE increases in the ILR image, affecting the curve shape

region where tissue is homogeneous, the noise-free curves do not differ. The noisy curves vary only due to the random noise.

A different result is observable for the curve sampled from a lesion. The noise-free curves are not identical in this case, because significant PVE occurs for the ILR volume. This has visible effect on the amount of enhancement and initial intensity value, as well as the slope of the curve in the initial wash-in and final wash-out phases. For actively enhancing lesion, noise is effectively smaller than for the mildly enhancing normal white matter as compared to signal variations due to enhancement.

## 4.2 Semi-quantitative DCE Analysis

In analysis of the simulated DCE datasets, we first performed semi-quantitative modelling. We estimated the time to peak and the mean and wash-out gradients for noise-free and noisy HR and ILR volumes, for all voxels were lesion tissue was dominant. Mean values and standard deviations of these estimates are presented in Table 2. The time to peak was slightly underestimated in the noise-free HR DCE. For other volumes, the value of this feature was overestimated, but the largest error and standard deviation is observed for the noisy ILR volume. For the wash-in gradient, its value was significantly underestimated for the noisy ILR DCE images, suggesting that is might be misleading to use it for diagnostic purposes. The wash-out gradient was underestimated for both HR and ILR images. The standard deviation of the latter was yet more than twice larger.

**Table 2.** Mean and standard deviation of the semi-quantitative pharmacokinetic parameters in voxels where MS lesion tissue is dominant. See Sect. 3.7 for equations

| Image volume | $t_3$ [min] | | $G_{in}$ [min$^{-1}$] | | $G_{out}$ [min$^{-1}$] | |
|---|---|---|---|---|---|---|
| | Mean | St. dev | Mean | St. dev | Mean | St. dev |
| HR | 1.28 | 0.07 | 7.31 | 1.75 | −0.13 | 0.03 |
| HR + 3% N | 1.60 | 0.43 | 4.12 | 2.80 | −0.17 | 0.05 |
| ILR | 1.61 | 0.30 | 3.44 | 2.27 | −0.06 | 0.04 |
| ILR + 3% N | 1.91 | 0.73 | 1.77 | 1.85 | −0.11 | 0.10 |
| MS lesion value | 1.3 | – | 11.37 | – | −0.20 | – |

## 4.3 Fully Quantitative DCE Analysis

Estimates of $K^{trans}$, $k_{ep}$ and $v_p$ were obtained and then averaged over all voxels when the lesion tissue was dominant. The results are presented in Table 3. Comparison to the ground truth values reveals that noisy ILR volumes, which are designed to be the most similar to actual DCE datasets, are associated with the largest standard deviation of the estimates. The errors in the mean values are also largest for the ILR images. It is worth noting that even the HR volumes resulted in estimates relatively far from the ground truth values.

**Table 3.** Mean and standard deviation of the quantitative pharmacokinetic parameters in voxels where MS lesion tissue is dominant. See Sect. 3.2 for explanation of the parameters

| Image volume | $K^{trans}$ $[10^{-3} \cdot \min^{-1}]$ | | $k_{ep}$ $[\min^{-1}]$ | | $v_p$ | |
|---|---|---|---|---|---|---|
| | Mean | St. dev | Mean | St. dev | Mean | St. dev |
| HR | 6.9 | 1.5 | 0.96 | 0.03 | 0.019 | 0.004 |
| HR + 3% N | 7.1 | 3.1 | 0.97 | 0.08 | 0.018 | 0.008 |
| ILR | 3.7 | 1.8 | 0.84 | 0.10 | 0.012 | 0.005 |
| ILR + 3% N | 3.5 | 2.5 | 0.86 | 0.20 | 0.009 | 0.015 |
| MS lesion value | 10 | – | 1 | – | 0.03 | – |

# 5  Discussion and Conclusions

The results clearly show the negative impact of the PVE on estimation of both semi- and fully-quantitative pharmacokinetic features in voxels where lesion tissue is dominant. Although we have simulated a brain dataset with multiple sclerosis, as allowed by the publicly available anatomical data, we think that similar results would be obtained for other organs and other lesion types. Therefore, the main point of our work is to acknowledge the opinions that spatial resolution is the crucial factor in DCE analysis, regardless of the analysis type. It is also clear that magnitude of errors is feature-dependent. In our analysis, time to peak appeared to be estimated with lower error than wash-in gradient, for example. The reason could be that averaging and downsampling, which is the source of spatial resolution loss in ILR DCE images, affect different parts of the time-intensity curves to a variable degree.

3% noise that was added to HR and LR images showed to have smaller impact on the estimates of fully quantitative parameters than 6 mm slice thickness. For semi-quantitative parameters, this effect was not so clear. The estimate of $t_3$ and $G_{in}$ was affected to a similar degree by noise and slice thickness. However, for the $G_{out}$ parameter, noise lead to overestimation of the negative slope, while slice thickness cause underestimation. Both noise and slice thickness increased the standard deviation of the estimate. This is understandable for random noise. For slice thickness, it implies that different voxels are affected to a variable degree by averaging and downsampling leading to thick slices.

Considering possible drawbacks of our study, we first note that our DCE simulations are based on imaging and perfusion models that do not reflect the complex phenomena of both image acquisition and pharmacokinetics perfectly. In our experiment, HR image acquisition was simulated using signal equations and LR image acquisition was modelled using averaging and downsampling. Both models simplify the complex nature of MR imaging and MR signal measurements and do not reflect all phenomena that occur during imaging. The pharmacokinetic model of Tofts-Kety that we used to simulate the CA kinetics is a popular, but again simplified description of the true, complex processes of

CA passage through vessels and tissues. The anatomical model of the brain is simplified as well, it does not include vessels, for example. Lesions were modelled as homogeneous regions, while in reality, they are heterogeneous. Nonetheless, we believe that the above simplifications do not hinder our goals of studying the influence of spatial resolution on DCE image analysis results, since this analysis always assumes some simplifications. Moreover, this influence might be common to various models, as it basically results from signal sampling theory. We find the simulated images useful to study the influence of such factors as noise or PVE on the biomarkers computed from the images. In the future, it is possible to use more complicated models of the imaging process and CA kinetics to improve the realisticness of the simulated DCE dataset. To the best of our knowledge, however, we are first to simulate DCE images with adjustable slice thickness, noise and such that incorporate the partial volume effect. Such images can be of great use in future studies, for example for assessment of super-resolution methods or pharmacokinetic modelling methods, since they allow a truly quantitative evaluation when kinetic feature values are known by design.

We have successfully simulated DCE MRI images using brain anatomy. We used them to study the impact of spatial resolution on pharmacokinetics-related biomarkers, which revealed the lacks of typical DCE images, which usually have considerable slice thickness.

## Acronyms

AIF - arterial input function,
BBB - blood-brain barrier,
CA - contrast agent,
DCE MRI - dynamic contrast-enhanced magnetic resonance imaging,
FA - flip angle,
Gd-DTPA - gadopentate dimeglumine,
HR - high resolution,
ILR - interpolated low resolution,
LR - low resolution,
PD - proton density,
PVE - partial volume effect,
TE - time to echo,
TR - repetition time,
T1W - T1-weighted,
VFA - variable flip angle,
WM - white matter

## References

1. Profile: DCE MRI quantification (2012). http://qibawiki.rsna.org/index.php/Profiles

2. Banerji, A.: Modelling and simulation of dynamic contrast-enhanced MRI of abdominal tumours. Ph.D. thesis (2012)
3. Betrouni, N., Tartare, G.: ProstateAtlas SimDCE: a simulation tool for dynamic contrast enhanced imaging of prostate. IRBM **36**(3), 166–169 (2015)
4. Bosca, R.J., Jackson, E.F.: Creating an anthropomorphic digital MR phantom-an extensible tool for comparing and evaluating quantitative imaging algorithms. Phys. Med. Biol. **61**(2), 974 (2016)
5. Collins, D.L., et al.: Design and construction of a realistic digital brain phantom. IEEE Trans. Med. Imaging **17**, 463–468 (1998). https://doi.org/10.1109/42.712135
6. Cuenod, C.A., Balvay, D.: Perfusion and vascular permeability: basic concepts and measurement in DCE-CT and DCE-MRI. Diagn. interv. Imaging **94**, 1187–1204 (2013). https://doi.org/10.1016/j.diii.2013.10.010
7. Dikaios, N., Arridge, S., Hamy, V., Punwani, S., Atkinson, D.: Direct parametric reconstruction from undersampled (k, t)-space data in dynamic contrast enhanced MRI. Med. Image Anal. **18**(7), 989–1001 (2014)
8. Fabijańska, A.: A novel approach for quantification of time-intensity curves in a DCE-MRI image series with an application to prostate cancer. Comput. Biol. Med. **73**, 119–130 (2016). https://doi.org/10.1016/j.compbiomed.2016.04.010
9. Furman-Haran, E., Grobgeld, D., Kelcz, F., Degani, H.: Critical role of spatial resolution in dynamic contrast-enhanced breast MRI. J. Magn. Reson. Imaging (JMRI) **13**, 862–867 (2001). https://doi.org/10.1002/jmri.1123
10. Gudbjartsson, H., Patz, S.: The Rician distribution of noisy MRI data. Magn. Reson. Med. **34**, 910–914 (1995). https://doi.org/10.1002/mrm.1910340618
11. Haq, N.F., Kozlowski, P., Jones, E.C., Chang, S.D., Goldenberg, S.L., Moradi, M.: A data-driven approach to prostate cancer detection from dynamic contrast enhanced MRI. Comput. Med. Imaging Graph. Off. J. Comput. Med. Imaging Soc. **41**, 37–45 (2015). https://doi.org/10.1016/j.compmedimag.2014.06.017
12. He, D., Xu, L., Qian, W., Clarke, J., Fan, X.: A simulation study comparing nine mathematical models of arterial input function for dynamic contrast enhanced MRI to the Parker model. Australas. Phys. Eng. Sci. Med. **41**(2), 507–518 (2018). https://doi.org/10.1007/s13246-018-0632-0
13. Jurek, J.: Super-resolution reconstruction of three dimensional magnetic resonance images using deep and transfer learning. Ph.D. thesis (2020)
14. Jurek, J., Kociński, M., Materka, A., Elgalal, M., Majos, A.: CNN-based superresolution reconstruction of 3D MR images using thick-slice scans. Biocybern. Biomed. Eng. **40**(1), 111–125 (2020)
15. Khalifa, F., et al.: Models and methods for analyzing DCE-MRI: a review. Med. Phys. **41**, 124301 (2014). https://doi.org/10.1118/1.4898202
16. Kwan, R.K., Evans, A.C., Pike, G.B.: MRI simulation-based evaluation of image-processing and classification methods. IEEE Trans. Med. Imaging **18**, 1085–1097 (1999). https://doi.org/10.1109/42.816072
17. Kwan, R.K.-S., Evans, A.C., Pike, G.B.: An extensible MRI simulator for post-processing evaluation. In: Höhne, K.H., Kikinis, R. (eds.) VBC 1996. LNCS, vol. 1131, pp. 135–140. Springer, Heidelberg (1996). https://doi.org/10.1007/BFb0046947
18. van der Leij, C., Lavini, C., van de Sande, M.G.H., de Hair, M.J.H., Wijffels, C., Maas, M.: Reproducibility of DCE-MRI time-intensity curve-shape analysis in patients with knee arthritis: a comparison with qualitative and pharmacokinetic analyses. J. Magn. Reson. Imaging (JMRI) **42**, 1497–1506 (2015). https://doi.org/10.1002/jmri.24933

19. O'Connor, J., Tofts, P., Miles, K., Parkes, L., Thompson, G., Jackson, A.: Dynamic contrast-enhanced imaging techniques: CT and MRI. Br. J. Radiol. **84**(special_issue_2), S112–S120 (2011)
20. Orton, M.R., et al.: Computationally efficient vascular input function models for quantitative kinetic modelling using DCE-MRI. Phys. Med. Biol. **53**, 1225–1239 (2008). https://doi.org/10.1088/0031-9155/53/5/005
21. Pannetier, N.A., Debacker, C.S., Mauconduit, F., Christen, T., Barbier, E.L.: A simulation tool for dynamic contrast enhanced MRI. PLoS ONE **8**, e57636 (2013). https://doi.org/10.1371/journal.pone.0057636
22. Parker, G.J.M., et al.: Experimentally-derived functional form for a population-averaged high-temporal-resolution arterial input function for dynamic contrast-enhanced MRI. Magn. Reson. Med. **56**, 993–1000 (2006). https://doi.org/10.1002/mrm.21066
23. Reichenbach, J., Hackländer, T., Harth, T., Hofer, M., Rassek, M., Mödder, U.: 1H T1 and T2 measurements of the MR imaging contrast agents Gd-DTPA and Gd-DTPA BMA at 1.5T. Eur. Radiol. **7**(2), 264–274 (1997). https://doi.org/10.1007/s003300050149
24. Tofts, P.S.: Modeling tracer kinetics in dynamic Gd-DTPA MR imaging. J. Magn. Reson. Imaging (JMRI) **7**, 91–101 (1997). https://doi.org/10.1002/jmri.1880070113
25. Tofts, P.S., Kermode, A.G.: Measurement of the blood-brain barrier permeability and leakage space using dynamic MR imaging. 1. Fundamental concepts. Mag. Reson. Med. **17**(2), 357–367 (1991)
26. Weinmann, H.J., Laniado, M., Mützel, W.: Pharmacokinetics of GdDTPA/dimeglumine after intravenous injection into healthy volunteers. Physiol. Chem. Phys. Med. NMR **16**(2), 167–172 (1984)
27. Yankeelov, T., Gore, J.: Dynamic contrast enhanced magnetic resonance imaging in oncology: theory, data acquisition, analysis, and examples. Curr. Med. Imaging Rev. **3**(2), 91–107 (2009). https://doi.org/10.2174/157340507780619179

# A New Sport Teams Logo Dataset
# for Detection Tasks

Andrey Kuznetsov[1(✉)] and Andrey V. Savchenko[1,2]

[1] St. Petersburg Department of Steklov Institute of Mathematics,
St. Petersburg, Russia
kuznetsoff.andrey@gmail.com
[2] Laboratory of Algorithms and Technologies for Network Analysis, National
Research University Higher School of Economics, Nizhny Novgorod, Russia
avsavchenko@hse.ru

**Abstract.** In this research we introduce a new labelled SportLogo dataset, that contains images of two kinds of sports: hockey (NHL) and basketball (NBA). This dataset presents several challenges typical for logo detection tasks. A huge number of occlusions and logo view changes during playing games lead to an ambiguity of a straightforward detection approach use. Another issue is logo style changes due to seasonal kits updates. In this paper we propose a two stage approach, in which, firstly, an input image is processed by a specially trained scene recognition convolutional neural network. Second, conventional object detectors are applied only for sport scenes. Experimental study contains results of different combinations of backbone and detector convolutional neural networks. It was shown that MobileNet + YOLO v3 solution provides the best quality results on the designed dataset (mAP = 0.74, Recall = 0.87).

**Keywords:** Logo detection · Sport logo dataset · Deep learning · Convolutional neural networks · Object detector.

## 1 Introduction

Object detection is of a high priority task for computer vision and machine learning. This problem has a wide range of applications in many domains [1–6]: market brand research for advertising planning, brand logo recognition for commercial research, vehicle logo recognition for artificial transport systems, etc. Some specific problems are solved with well-known approaches that use SIFT or SURF keypoint detectors and local descriptors based on them. In ordinary object detection tasks deep learning methods are frequently used and are quite successful in terms of quality results [7–9]. Deep learning object detection model usually requires many labelled training data [10,11]. In some tasks a dataset augmentation helps to increase the training set volume [12]. However, extensive datasets are not necessarily available in many cases such as logo detection where the publicly available datasets are very small (Table 1) and mostly oriented on brand logos and web-based imagery. Small datasets are less applicable to deep

L. J. Chmielewski et al. (Eds.): ICCVG 2020, LNCS 12334, pp. 87–97, 2020.
https://doi.org/10.1007/978-3-030-59006-2_8

learning tasks, so keypoint based seem to be more useful for this kind of detection problem. That is why in our previous research [13] we started developing keypoint-based methods to check their quality.

**Table 1.** Existing logo detection datasets

| Dataset | Objects | Images | Open access |
|---|---|---|---|
| FlickrLogos-27 [14] | 27 | 1080 | Yes |
| FlickrLogos-32 [1] | 32 | 2240 | Yes |
| BelgaLogos [13] | 37 | 1951 | Yes |
| LOGO-Net | 160 | 73414 | No |
| SportLogo | 31 | 2836 | Yes |

Thus, the main contribution of this paper is a novel dataset of sport logos. In contrast to all existing attempts, we collected and labelled for National Basketball Association (NBA) and National Hockey League (NHL) teams and made it publicly available[1]. The labelled SportLogo dataset can be easily augmented on the training stage not to suffer from insufficient data volume. We also provide a simple scheme for extending the SportLogo dataset with new sport logos if needed. Moreover, we examine several well-known object detectors and provide the best methods. We investigate the deep learning approaches based on YOLO v3 and SSD detector architectures and made a comparison of different combinations of backbone feature extraction and detector architectures. In addition, we propose a two-stage approach that classifies scenes in the input image and execute object detector only for sport scenes.

It should be mentioned that logo detection algorithms are not limited by the tasks of logos. The developed methods can be adopted for traffic signs detection, medical markers detection, cancer recognition, etc. Obviously, the proposed approach can be used in different research areas, which are oriented on a similar problem analysis - to detect small significant image regions which can be transformed differently due to an unconstrained environment.

## 2 Related Work

Most of the works on logo detection describe the methods and algorithms that use small scales, in both the number of logo images and logo classes [5] due to the high costs in creating large scale logo datasets. It is rather hard to collect automatically large-scale logo training dataset that covers many different logo classes. Web data mining can be applied for this problem solution as shown in some other studies [9]. Nevertheless, it is difficult to acquire accurate logo annotations since no bounding box annotation is available from typical web images. To solve this problem, we used specific sport logo web resources where NHL and NBA teams' logos can be downloaded.

---

[1] https://github.com/kuznetsoffandrey/SportLogo.

Most approaches to logo detection are based on local features calculation, e.g., SIFT, color histogram, edge [1–5]. They can obtain more detailed representation and provide model robustness for recognizing a large number of different logos with less training complexity. One of the reasons is insufficient labelled datasets that are required for developing deep solutions. For example, among all publicly available logo datasets, the most common FlickrLogos-32 dataset [1] contains 32 logo classes each with only 70 images and in total 5644 logo objects, BelgaLogos [14] has 2695 logo images from 37 logo classes with bounding box labelled annotations (Table 1).

However, a few deep learning approaches for logo detection have been published recently. Iandola et al. [15] applied Fast R-CNN model for logo detection, which hardly suffers from the small training data size. To facilitate deep learning logo detection, Hoi et al. [16] built a large-scale dataset called LOGO-Net by collecting images from online retailer websites and then manually label them. This requires a huge amount of construction effort and moreover, LOGO-Net is inaccessible publicly (see Table 1).

In our previous study we preliminary explored the potentials for many existing keypoint-based algorithms for sport logo detection [13]. Unfortunately, the keypoint-based solutions could not detect very small logos comparing to the image size. Thus, in this paper in addition to the dataset itself we propose the novel approach, that shows high efficiency in an unconstrained environment when team logos are transformed in a very complex way and extremely occluded in some cases.

## 3 SportLogo Dataset

Due to the absence of a sport logos dataset we decided to create the dataset by ourselves. We planned to add several popular kinds of sports in this dataset and label the corresponding images. As for the dataset creation process, we took NBA and NHL teams (the list of NHL and NBA teams has been taken from the official web sites). The next step of our dataset creation was providing an image download tool that could use the list of team names to find relative images. Using this tool, we could download approximately 100 images for every sport team and labelled a part of the dataset that contains NHL and NBA teams.

During investigation process we found a Github repository for downloading images using Google search engine (https://github.com/hardikvasa/google-images-download) which is able to download images by keywords or key images taking into account different search parameters. Further the list of the main parameters is presented with a short description:

– *keywords* - denotes the keywords/key phrases you want to search for. We use team title and sometimes add a game season year if logo has changed;
– *format* - denotes the format/extension of the image that you want to download (jpg, gif, png, bmp, svg, webp, ico). We use jpg and png images;
– *specific_site* - allows to download images with keywords only from a specific website/domain name you mention. We use it to download logo images that we try to find a match for in the dataset images.

Using the parameters above we were able to download images with specific search settings. To simplify the downloading process a script has been developed that uses input text file with keywords (one per line), creates the output folder and downloads images to it for every NHL and NBA team. As a result, we downloaded 3100 images for 31 NHL teams and 3000 images for NBA teams. At the end of the download process we made a closer look to the downloaded images and removed some irrelevant images that contained no team logos - as a result 5500 relevant images were selected in total. The total amount of data collected is takes nearly 900 Mb drive space. The dataset was divided into three parts: train, test and validation. Train/test set contains 2819 images and corresponding text files with labels in VOC style *(train_test_data* folder), the validation set contains 151 images (*val_data* folder) for 61 teams. Team names are presented in *classes.txt* file.

The examples of SportLogo images are presented in Fig. 1. It should be mentioned that the dataset has been fully labelled with bounding boxes of logos positions.

(a)

(b)

**Fig. 1.** Sample images from SportLogo dataset: (a) NHL; (b) NBA

# 4  Sport Logo Detection Techniques

## 4.1  Keypoint-Based Solutions

As it was stated above, we first tried to develop a feature-based logo detection algorithm and provided the comparison of different solutions [13]. The following pipeline was developed to detect sport logos:

1. Load logos images and test images;
2. Calculate descriptors using one of selected algorithms (SIFT, SURF, ORB, BRISK, FREAK, AKAZE or deep local features DELF [17,18]);
3. Use FLANN (Fast Library for Approximate Nearest Neighbors) matcher or BF (Brute Force) matcher to find matching descriptors, evaluated on the previous step;
4. Estimate a list of good matches using a distance threshold defined for every extraction algorithm;
5. In case the good matches number is enough to make decision about the correctness of a match, then apply RANSAC algorithm to find homography among keypoints left in good matches list;
6. Based on found inliers make decision if the logo presents on a test image or not.

The quantitative and qualitative results presented in previous research [13] showed that SIFT and SURF descriptors calculation methods show better or comparable results with the existing configuration of state-of-art deep learning based DELF algorithm. However, this keypoint-based solution is very limited by logos transformations and quite hard to be applied in real life scenarios when NHL or NBA games photos should be recognized.

## 4.2  Deep Learning-Based Approach

During previous research we tried different keypoint-based approaches for sport logo detection. The best results we could achieve were obtained on a small dataset with almost clear NHL logos without occlusion or any difference in logo design (a lot of images of sport logos exist that are differently designed: created in image processing software, stitched on clothes, simplified logos, etc.). Due to this fact we decided to use object detection CNNs for logo detection. During analysis of state-of-the-art network architectures and open source solutions we found several existing logo detection solutions. A good example is the DeepLogo system which is in fact a brand logo detection software that is based on region-based convolutional neural networks. The authors tested it on several brand logos detection. Moreover, they used augmentation to make the training set larger - they used 4000 of images from Flickr dataset and obtained approximately 140000 images for training. DeepLogo CNN is based on CAPTCHA detection and recognition network.

We tried to train this network on the labeled Sport Logos dataset but did not reach the goal and did not achieve satisfying results. So other networks had to be analyzed. We took the most popular detectors: vanilla YOLO and SSD300 and started training. The results of comparison will be given in Subsection 3.6.2. During experimental research we also defined different combinations of detectors (YOLO v3, tiny YOLO v3, SSD7, SSD300, SSD512) and feature extraction backbone networks (VGG16, MobileNet, MobileNet v2). Moreover, we analyzed YOLO training mechanism using Darknet and Keras engines. As it will be shown further, the best combination defined was MobileNet + YOLO v3. This combination provides the best results on the designed SportLogo dataset with occlusions and other hard cases. Several results of sport logo detections are presented in Fig. 2.

**Fig. 2.** SportLogo detections using YOLO v3 detector architecture

The proposed pipeline uses obtained results on scene detection CNN [19, 20] trained on Places2 dataset. The following pipeline contains of two major steps (also illustrated in Fig. 3):

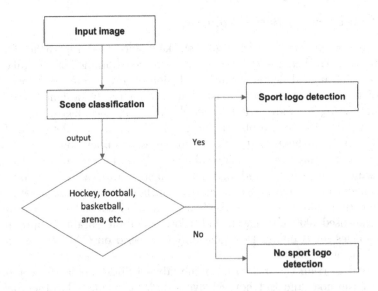

**Fig. 3.** Image processing pipeline for sport logo detection.

1. Scene classification to estimate if the image contains sports in any form. We selected several predicted classes to obtain better scene filtering, because we do not need to analyze non sport images for sport logo detection. Of all the dataset 90% images were correctly classified as containing sport scenes. The group of sport labels includes hockey, football, basketball, arena and will be extended. The list of classes will be extended in the nearest further. This step leads to sport logo detection speed improvement (we analyze only specific images).
2. Sport logo detection step based on feature extraction with CNNs is applied only for sport images and uses the best combination of detector and classifier networks we found during research: MobileNet and YOLO v3.

## 5   Experimental Research

First, we calculated scene classification results in a separate table to estimate the accuracy of the first step of the proposed pipeline. The results are presented in Table 2. It should be mentioned that we can select a subspace of scene classification outputs that can be used as a unique group for sport scene classification: 'hockey', 'football', 'basketball court', 'arena (performance) and 'locker room'. Obviously, this group should be extended further with adding new sports and changing classes structure. Using this group of output labels for sport images classification we achieve 94% and 90% accuracy for NHL and NBA images correspondingly (here by accuracy we mean the number of correct sport scenes classification divided by the total number of sport scenes in the validation set for NHL and NBA images).

**Table 2.** Sport logo scenes classification results for NHL and NBA images (the first step of the proposed pipeline).

| Predicted scene | NHL data (148 images) | NBA data (57 images) |
| --- | --- | --- |
| Hockey | 115 | 2 |
| Football | 15 | 19 |
| Arena (performance) | 3 | 10 |
| Locker room | 6 | 1 |
| Basketball court | 0 | 19 |
| Other | 9 | 6 |

We created a validation dataset for 61 teams (4 teams among them have high similarity of fonts on the kits and as a result - high misclassification rate), that include NHL and NBA teams with distinguishable logos. It should be mentioned that if NHL teams have specific team logos on their kits, NBA teams do not have them, moreover NBA kits are much more varying in comparison to NHL) The learning is implemented with 2 images in a mini-batch, dynamic learning rate depending on the number of epochs and a combination of Adam and SGD optimizers (90 epochs with Adam, 10 - with SGD). The validation dataset contains images, downloaded for all the teams specifically and does not intersect with train/test sets used for MobileNet + YOLO v3 training. This dataset contains 151 images. The best results of features and detectors combinations are presented in Table 3. We analyzed not only descriptors and backbone networks combinations, but also the frameworks used during the training process. Speaking about backbone feature extraction networks we used VGG16, MobileNet and MobileNet v2. We used SSD7, SSD300, YOLO v3 (both full and tiny architectures implemented on Keras and Darknet) on detection step. The last one (YOLO v3) was used for different input sizes for the inference. As we can see from Table 3 the highest quality values in terms of mAP and Recall metrics are obtained for MobileNet + YOLO v3 with input image size scaled to $480 \times 480$: mAP $= 74.03$, Recall $= 0.87$.

**Table 3.** Comparison of various object detectors for the SportLogo dataset.

| Backbone | Detector | mAP,% | Recall,% |
|---|---|---|---|
| VGG16 (COCO) | SSD300 | 42.91 | 44.12 |
| VGG16 | SSD300 | 28.21 | 36.12 |
| VGG16 | SSD7 | 45.07 | 52.92 |
| VGG16 | YOLO v3 (320) | 59.99 | 63.83 |
| VGG16 | YOLO v3 (320) - Darknet | 43.50 | 44.34 |
| VGG16 | YOLO v3 tiny (320) - Darknet | 60.35 | 62.02 |
| MobileNet | YOLO v3 (320) | 68.20 | 69.86 |
| MobileNet | YOLO v3 (480) | 74.03 | 87.12 |
| MobileNet | YOLO v3 (960) | 58.58 | 61.43 |
| MobileNet | SSD300 | 34.17 | 33.34 |
| MobileNet v2 | SSD300 | 36.66 | 37.78 |

We also did research on training the best combination of backbone/detector MobileNet + YOLO v3 for different kinds of sports separately with the number of classes equals to 31 and 30 for NHL and NBA respectively. Experiments shown an improvement in quality results (see Table 4). In this table we also show the difference in accuracy results for separate models and the merged model. We can see that when using specific datasets for separate kinds of sports the merged dataset shows slightly worse results for NHL part of the dataset (1–2% difference), whereas for NBA the difference is bigger (5–7%). This experiments also show what an impact is made by NBA teams which do not have clear logos on their kits. However, even due to quality improvements which appear if we use separate models, this will also lead to significant speed decrease, so a combined

**Table 4.** Cross-dataset results of MobileNet+YOLO v3 for the SportLogo dataset.

| Training set | Validation set | mAP, % | Recall,% |
|---|---|---|---|
| NHL | NHL | 90 | 98 |
| NHL + NBA | NHL | 89 | 97 |
| NHL | NBA | 66 | 79 |
| NHL + NBA | NBA | 56 | 74 |
| NHL + NBA | NHL + NBA | 74 | 87 |

NHL+NBA model seems be the best solution for sport logo detection on a mobile platform.

Some of the qualitative results are presented in Fig. 4 for several sport scenes with single and several teams logos detection.

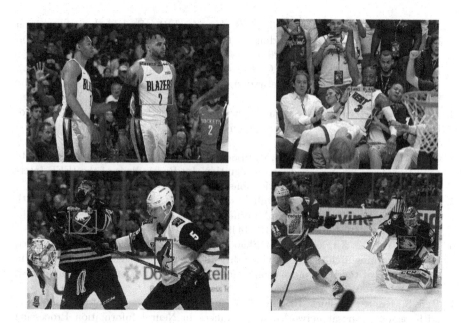

**Fig. 4.** Sample results of NHL/NBA team's logos recognition.

## 6  Conclusion

Though our sport logo detectors for NHL and NBA teams show high accuracy, they may be outperformed if the human-labeled segmentation is used instead of bounding boxes of logos or kit parts. It can be used for specific kinds of sports like NBA and football, where different logos are placed on a kit (e.g., sponsor names). However, the sport logo detection network should be fine-tuned if new logos should be detected or the kit of a team changes from season to season. Despite this the proposed deep learning approach showed high quality metrics values especially in comparison with keypoint-based approaches (increase by 20%).

We also compared different combinations of feature extraction and detection convolutional neural networks to select the best pair: MobileNet and YOLO v3. We also provided a pipeline for sport logo detection based on scene classification and further sport logo detection. We could reach the quality goal using the proposed solution: mAP=0.74 and Recall=0.87. The average logo detection speed on GeForce GTC 1070 takes 0.05 s, whereas on CPU it takes 0.3-0.5 s.

**Acknowledgements.** This research is based on the work supported by Samsung Research, Samsung Electronics.

# References

1. Romberg, S., Pueyo, L. G., Lienhart, R., Van Zwol, R. : Scalable logo recognition in real-world images. In: Proceedings of the 1st ACM International Conference on Multimedia Retrieval, vol. 25 (2011)
2. Revaud, J., Douze, M., Schmid, C.: Correlation-based burstiness for logo retrieval. In: Proceedings of the ACM International Conference on Multimedia, pp. 965–968 (2012)
3. Romberg, S., Lienhart, R. : Bundle min-hashing for logo recognition. In: Proceedings of the 3rd ACM Conference on International Conference on Multimedia Retrieval, pp. 113–120 (2013)
4. Boia, R., Bandrabur, A., Florea, C. : Local description using multi-scale complete rank transform for improved logo recognition. In: Proceedings of the IEEE International Conference on Communications, pp. 1–4 (2014)
5. Li, K.-W., Chen, S.-Y., Su, S., Duh, D.-J., Zhang, H., Li, S.: Logo detection with extendibility and discrimination. Multimedia Tools Appl. **72**(2), 1285–1310 (2013). https://doi.org/10.1007/s11042-013-1449-1
6. Pan, C., Yan, Z., Xu, X., Sun, M., Shao, J, Wu, D.: Vehicle logo recognition based on deep learning architecture in video surveillance for intelligent traffic system. In: Proceedings of the IET International Conference on Smart and Sustainable City, pp. 123–126 (2013)
7. Ren, S., He, K., Girshick, R., Sun, J.: Faster R-CNN: towards real-time detection with region proposal networks. In: Advances in Neural Information Processing Systems, pp. 91–99 (2015)
8. Redmon, J., Divvala, S., Girshick, R., Farhadi, A.: You only look once: unified, real-time object detection. In: Proceedings of the IEEE Conference on Computer Vision and Pattern Recognition, pp. 779–788 (2016)
9. Shrivastava, A., Gupta, A., Girshick, R.: Training region-based object detectors with online hard example mining. arXiv preprint arXiv:1604.03540 (2016)
10. Simonyan, K., Zisserman, A.: Very deep convolutional networks for large-scale image recognition (2014). arXiv preprint arXiv:1409.1556
11. Krizhevsky, A., Sutskever, I., Hinton, G. E.: ImageNet classification with deep convolutional neural networks. In: Advances in Neural Information Processing Systems, pp. 1097–1105 (2012)

12. Shorten, C., Khoshgoftaar, T.M.: A survey on image data augmentation for deep learning. J. Big Data **6**, 60 (2019)
13. Kuznetsov, A., Savchenko, A.: Sport teams logo detection based on deep local features. In: Proceedings of IEEE International Multi-Conference on Engineering, Computer and Information Sciences (SIBIRCON), pp. 0548–0552 (2019)
14. Joly, A., Buisson, O.: Logo retrieval with a contrario visual query expansion. In: Proceedings of the ACM International Conference on Multimedia, pp. 581–584 (2009)
15. Iandola, F. N., Shen, A, Gao, P., Keutzer, K.: DeepLogo: hitting logo recognition with the deep neural network hammer. arXiv preprint arXiv:1510.0213 (2015)
16. Hoi, S.C., et al.: Logo-Net: large-scale deep logo detection and brand recognition with deep region-based convolutional networks. arXiv preprint arXiv:1511.02462 (2015)
17. Noh, H., Araujo, A., Sim, J., Weyand, T., Han, B.: Large-scale image retrieval with attentive deep local features. In: Proceedings of the International Conference on Computer Vision, ICCV, pp. 3476–3485 (2017)
18. Kuznetsov, A.: Face recognition using DELF feature descriptors on RGB-D data. In: Proceedings of the International Conference - Analysis of Images, Social Networks and Texts (AIST), CCIS 1086, pp. 237–243 (2019)
19. Savchenko, A.V., Rassadin, A.G.: Scene recognition in user preference prediction based on classification of deep embeddings and object detection. In: Lu, H., Tang, H., Wang, Z. (eds.) ISNN 2019. LNCS, vol. 11555, pp. 422–430. Springer, Cham (2019). https://doi.org/10.1007/978-3-030-22808-8_41
20. Savchenko, A.V., Demochkin, K.V., Grechikhin, I.S.: User preference prediction in visual data on mobile devices. arXiv preprint arXiv:1907.04519 (2019)

# Scene Recognition Using AlexNet to Recognize Significant Events Within Cricket Game Footage

Tevin Moodley⬤ and Dustin van der Haar$^{(\boxtimes)}$⬤

Academy of Computer Science and Software Engineering,
University of Johannesburg, Kingsway Avenue and, University Rd,
Johannesburg 2092, South Africa
{tevin,dvanderhaar}@uj.ac.za

**Abstract.** In the last decade, special attention has been made toward the automated analysis of human activity and other related fields. Cricket, as a research field, has of late received more attention due to its increased popularity. The cricket domain currently lacks datasets, specifically relating to cricket strokes. The limited datasets restrict the amount of research within the environment. In the study, this research paper proposes a scene recognition model to recognize frames with a cricket batsman. Two different classes are addressed, namely; the gameplay class and the stroke class. Two pipelines were evaluated; the first pipeline proposes the Support Vector Machine (SVM) algorithm, which undergoes data capturing, feature extraction using histogram of oriented gradients and lastly classification. The Support Vector Machine (SVM) model yielded an accuracy of 95.441%. The second pipeline is the AlexNet Convolutional Neural Network (CNN) architecture, which underwent data capturing, data augmentation that includes rescaling and shear zoom followed by feature extraction and classification using AlexNet. The AlexNet architecture performed exceptionally well, producing a model accuracy of 96.661%. The AlexNet pipeline is preferred over the Support Vector Machine pipeline for the domain. By recognizing a significant event, that is when a stroke and none stoke (gameplay) scene is recognized. The model is able to filter only relevant footage from large volumes of data, which is then later used for analysis. The research proves there is value in exploring deep-learning methods for scene recognition.

**Keywords:** Scene recognition · Automation · Cricket strokes · AlexNet architecture · Support vector machines

## 1 Introduction

Research in recent decades has paid special attention to application in robotics, information science, and automation and its important fields of human activity,

© Springer Nature Switzerland AG 2020
L. J. Chmielewski et al. (Eds.): ICCVG 2020, LNCS 12334, pp. 98–109, 2020.
https://doi.org/10.1007/978-3-030-59006-2_9

in particular, sport and culture [14]. However, despite the many years of extensive research, the problem of identifying significant scenes in a video depicting sporting events efficiently and effectively remains a challenge [11]. Cricket is a sport played by two teams, each consisting of eleven players each. Figure 1 depicts the inner workings of the game where each team is allowed to bat and bowl, and the objective is to score as many runs possible during each team's batting innings [12]. Many significant events impact the outcome of a cricket game. One such event occurs when the batsman is striking the ball. A possible reason may be due to the need to accumulate runs. The batsmen are required to strike the ball toward different areas of the field, thereby allowing maximum opportunity to accumulate runs for their respective teams. If the bowler dismisses a batsman, they are required to leave the field, signalling the end of their innings. Accumulating runs places emphasis on capturing the significant scene in a given match.

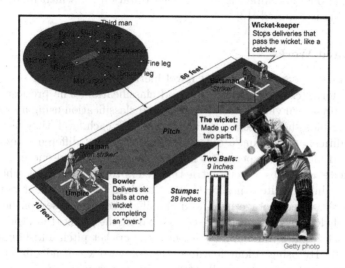

**Fig. 1.** Representation of how the game of cricket is played [1].

This research paper will discuss two solutions that are used to recognize significant events within a cricket game. Section 2 will focus on the problems and current works that exist within the domain. Section 3 will discuss the experiment setup, which will illustrate the manner in which the solutions were implemented in Sects. 4 and 5. Finally, the discussion in Sect. 7 will highlight the results in Sect. 6 of each pipeline with an emphasis on the performance achieved by the different algorithms.

## 2  Problem Background

Fei-Fei and Li note that one of the most remarkable features of the human visual system is the rate at which it can accurately and comprehensively recognize and interpret the complex visual world [4]. Understanding what we see and recognizing significant scenes has grown as digital video becomes more pervasive. As a result, the efficient ways of mining information from video data has become increasingly important [13]. Video in itself contains large amounts of data and has complexities that contribute to the difficulties when performing automated analysis [13]. In general, systems for category recognition on images and video use machine learning based on image descriptions to distinguish objects and scene categories [18]. However, recent works have seen a shift towards more deep learning-based methods.

Kolekar and Palaniappan state that cricket is one of the most popular sports in the world that is played across 17 different countries [13]. However, it is noted that less research has been reported on cricket as opposed to other sports such as soccer, tennis, basketball, and baseball. Possible reasons may be due to the increased complexity of the game. Due to cricket being played over more extended periods, an emphasis on the long duration for which highly efficient video pruning is required [13].

## 2.1 Related Works

Kolekar and Palaniappan present a novel hierarchical framework and practical algorithms for cricket event detection and classification. The proposed scheme performs a top-down video event detection and classification using a hierarchical tree which avoids shot detection and clustering. At each level, they use different data modalities, each having a different focus to extract different events, which is segmented into audio, spectators and the field view. Level one uses audio to extract excitement clips, where extraction refers to isolating clips whereby the spectators exhibit excitement. Level two classifies the excitement clips. Level 3 analyses the video footage itself and classifies the clips based on field views. Within the field views, the grass is used to separate the players from the audience. Finally, level 4 and 5 differentiate between the cricket pitch and boundaries of the field [13]. By using colour features, the framework can adequately classify players from spectators with recall scores raging between 85%-89% and precision scores between 83%-88%. While promising results are presented, their research highlights the difficulty in determining how significant an event is in relation to the game.

Current work conducted by Huggi and Nandyal identified the umpire actions as significant events since the umpire is responsible for making important decisions within a cricket game [9]. Their novel technique for recognition of scenes in cricket recording utilizes umpire hand flags or motions for the recognition of significant events. At that point, umpire edges are recognized from every scene and dissected utilizing both vertical, and even power projected profiles. The profiles relating to the umpire can be used as components for preparing the Random Forest (RF) classifier. They were able to achieve an 86% accuracy in distinguishing different signals made by the umpire.

Umpires are responsible for making important decisions in a game, such as the dismissal of a batsman, the signalling of illegal deliveries, the number of runs awarded for a given stroke, and more. However, recognizing umpire signals is limited and addressing other scenes may potentially provide more value. While promising work is shown, further contributions can be made in the way of recognizing scenes in which strokes are performed. A batsman is responsible for stroking the ball to different areas of the field to accumulate runs, thereby increasing their team's chance of winning the game. Therefore, detecting when a stroke is performed may potentially have a more significant impact within the domain, since performing strokes allow the accumulation of runs, which impacts on the outcome of the game. Addressing the recognition of strokes may help aid in the determination of an interesting event.

Scene recognition has been applied to assist in specific tasks within the cricket domain. Currently, it is used to recognize different gestures made by an umpire, but there is potential to do more. Despite the many successes within the scene recognition research field [9,13,13], many existing techniques rely on complex and demanding visual features, which make the techniques expensive to use on large volumes of archived content [11]. Cricket video footage contains large volumes of content as the game is played in different formats. Either the shorter formats of 20/50-over matches or the longer format known as test matches, which are played over 4 or 5 consecutive days [12]. With a vast number of playing hours, cricket is one of the sports with the most extended playing times [12]. Therefore, the need to recognize significant events, more notably when a batsman performs a stroke, the tracking of the ball, and the audiences, must be investigated. By recognizing scenes where strokes are performed, significant scenes may be captured and consequently allows one to process the large volumes of content within video footage.

## 3   Dataset and Experiment Setup

To develop the scene recognition model, the experiment setup consisted of different phases that include data capturing, data augmentation, feature extraction, and classification. The different phases will be used to create two pipelines that are evaluated to determine the most suitable solution. The first pipeline referred to as the Support Vector Machines (SVMs) pipeline, has a data capturing phase, a feature extraction phase that is achieved using histogram of orientated gradients (HOG), and a classification phase using the SVM algorithm. The second pipeline is the CNN AlexNet pipeline, which has the following process; data capturing, data augmentation using rescaling and shear zoom, and classification.

Within a typical supervised-learning based classification problem, the goal usually involves separating data into testing and training sets. Each instance on the training set contains one target value, which in this context is the class label and several attributes defining that label [8]. Therefore, the data capturing phase comprises of highlight video footage of various International Test matches that were saved to the local disk. Once the video footage was captured, an image dataset was compiled that consisted of two classes; the gameplay class and the strokes class. The strokes class is selected, as it is identified as a significant event within the game. The gameplay class is chosen to signal all other events that occur in a game to isolate the strokes class further. The class labelled strokes consisted of all scenes where strokes were performed, and the class labelled gameplay contained all other scenes within the cricket match. The frames extracted are then visually inspected, manually labelled, and placed into their respective classes. The process of the manual sorting required a vast amount of time, where videos that ranged between 8–10 min in length required an estimated combined time of 4 h. The lengthy process further supports on the research problem of automated analysis and contributes to the need for a model that can recognize when a stroke is performed.

A total of 7480 images is captured that were split into training and testing sets using a 90:10 rule. The 90:10 split translated into 6800 images belonging to a training set and 680 images belonging to the testing set. Within the training dataset, 3400 images were allocated to the stroke class and the 3400 images to the gameplay class. The testing dataset was also split equally with 340 images belonging to the stroke class and 340 images to the gameplay class. Through experimentation, the image dimensions were set to 100 (*width*) × 150 (*height*) as it was noted that these dimensions yielded the most notable results while requiring the least amount of computational complexity.

Within the convolutional neural network (CNN) pipeline, the AlexNet architecture was applied. The architecture consisted of max-pooling, a fully connected layer, the ReLu activation layer, and a softmax layer as recommended when applying the architecture [5].

## 4    HoG-SVM Pipeline

### 4.1    Feature Extraction

Recent works have demonstrated how the histogram of orientated gradients (HOGs) has been used to obtain efficient and robust shape-based cues in images [17,20]. The method is able to count the number of occurrences in localized portions of an image. Computation of local histograms of gradients is achieved by establishing the gradients within an image that will allow each cell to form a histogram of orientation, followed by normalization of each cell. The first step to compute the gradient of an image is to normalize the colour and gamma values. HOG achieves normalization by filtering the image with one-dimensional filters on the horizontal and vertical planes. The method requires colour filtering or the intensity of the data with the [−1 0 1] kernel, and the matrix transpose [17].

The proceeding step is referred to as cell and block descriptors, where the image is split into different cells. A cell is defined as a spatial region like a square with a predefined size in pixels [17]. For each cell, the histogram of orientation is computed by accumulating votes into bins for each orientation. Once the feature extraction phase is completed, the features are fed into a classifier that will attempt to make distinctions between the features.

## 4.2  Classification

To determine which parameters best suit the SVM classifier, a hyperparameter tuning study was conducted. Using Grid Search, the following parameters were inspected, regularization value, kernel, and gamma value. The regularization value, which acts as the control for the fitting parameters, where if the magnitude of the fitting parameters increases, the resulting cost function is penalized. The kernel, which is the convolution matrix used for image processing. Lastly, the gamma parameter, which is used to encode and decode the luminance values in image or video systems. The optimal parameters discovered were, a regularization value of 10, the kernel set to RBF, and the gamma value of 0.001. The most noteworthy parameter was the kernel, which had the most significant impact on the results, both the linear and Radial basis function (RBF) kernels were tested. The RBF kernel yielded better results, as it encapsulates the decision space boundary better.

## 5  AlexNet Pipeline

### 5.1  Data Augmentation

The images were scaled using 1./255, a shear range that is the angle in a counterclockwise direction in degrees, and a zoom range of 0.2, which specifies the random zoom for each image. The dataset undergoes the next step in the pipeline, which is feature extraction followed by classification.

### 5.2  Classification

The AlexNet architecture can be described with the first convolutional layer filters $N \times 227 \times 277 \times 3$ with 96 kernels, where the kernel size is $11 \times 11 \times 3$ size with a 4-pixel stride. AlexNet has LNR (local response normalization) that connects the layers using a spatial organized pattern [19]. The second convolutional layer has 256 kernels, where its kernel size is $5 \times 5 \times 48$. The third and fourth layers both have 384 kernels where the third layers kernel size is $3 \times 3 \times 256$, and the forth kernel's size is $3 \times 3 \times 192$. The fifth convolutional layer has 256 kernels, and its size is $3 \times 3 \times 192$. The first two fully-connected layers have 4096 neurons, and the last fully-connected layer is divided into the categories representing the two labelled classes in this study [19]. The activation layer used was ReLu, which is a rectified linear unit that returns the element-wise activation function. ReLu is simple, fast, and empirically seems to work well, in which some cases, it tends to converge much faster and reliably [16].

The output activation layer used is the softmax activation function, which is often used to map the non-normalized output of a network to a probability distribution over predicted classes. The Adam optimizer was chosen, as the Adam optimizer combines the best properties of AdaGrad and RMSProp to provide an optimization algorithm that is preferred when handling sparse gradients on noisy data. The main reason for the use of the AlexNet architecture is due to the fact the network is slightly deeper than LeNet with more filters per layer and with stacked convolutional layers. However, future research will investigate deeper networks to address the impact more layers has on the classification accuracy.

## 6    Results

Having a model that can recognize when a stroke is performed may potentially allow for the automation of the entire manual process (shown in Sect. 3) conducted in this research. The model will be able to recognize significant scenes, thereby addressing the problem, which relates to filtering key events in a cricket game footage. Whereby only significant scenes relating to strokes are highlighted [13].

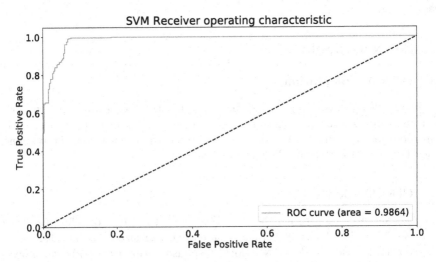

**Fig. 2.** The ROC for the SVM algorithm that achieves an area under the curve of 98.64%.

The classifiers are evaluated using Receiver Operating Characteristics (ROC) curves, the confusion matrix, accuracy, precision, recall, and the f1-score. The model accuracy, which will determine the ratio of correctly predicted observations to the total observations. The precision refers to the number of observations that are, in fact, true. Recall is the number of correctly predicted positive observations. Finally, f1-score is the weighted average between precision and recall [6].

Figure 2 is the ROC curve for the SVM algorithm that is able to summarize the trade-off between the true positive rate and false positive rate for a predictive model using different probability thresholds [15]. ROC curves provide an in-depth understanding of false positives [3]. Both ROC curves in Fig. 2 and Fig. 3 illustrate the trade-off between the different classes; gameplay and stroke. Achieving an SVM area under the curve of 98.64% is very promising, as it indicates the model's ability to identify the distinctions between actual positive and negative observations correctly. Further motivations will be outlined in the discussion section.

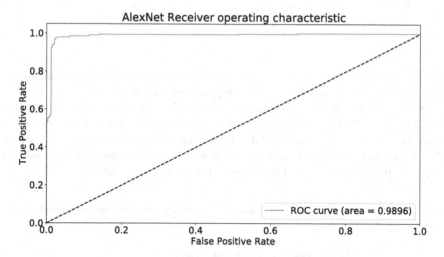

**Fig. 3.** The ROC for the AlexNet algorithm that achieves an area under the curve of 98.96%.

Figure 3 is the ROC graph representing the AlexNet architecture. The model produces a model of 96.661%, which demonstrates the exceptional performance of the architecture within the domain. With an AlexNet area under the curve of 98.96%, the model can distinguish between false positive and true positive observations adequately.

**Table 1.** The confusion matrix of the SVM classifier with a total of 31 false positives.

| Class | Actual gameplay class | Actual strokes class |
|---|---|---|
| Predicted gameplay class | 318 | 22 |
| Predicted strokes class | 9 | 331 |

The confusion matrices in Tables 1 and 2, highlight specific images that were falsely predicted as positive observations are identified. The false positive observations allowed the study to gauge how accurate each algorithm was in relation to one another. During an investigation, it is noted that the SVM algorithm produced a total of 22 false positives for the gameplay class and 9 false positives for the stroke class. The AlexNet architecture outperformed the SVM algorithm as it produced 16 false positives for the gameplay class and 7 for the stroke class.

**Table 2.** The confusion matrix of the AlexNet architecture with a total of 23 false positives.

| Class | Actual gameplay class | Actual strokes class |
|---|---|---|
| Predicted gameplay class | 324 | 16 |
| Predicted strokes class | 7 | 333 |

AlexNet produces fewer false positives than that of the SVM algorithm, which gives an early indication that it may be preferred for the domain.

From Table 3, accuracy, precision, recall, and f1-score weighted average scores are computed. From the results, the AlexNet architecture outperforms the SVM algorithm, which may be supported using the recall score, where AlexNet scores 97% as opposed to SVM's lower 95%. Further motivations will be made to demonstrate the AlexNet architecture's success within the domain.

**Table 3.** Performance comparison between the different algorithms conducted in this study.

| Metrics used for performance comparison | | | | |
|---|---|---|---|---|
| Algorithm | Accuracy | Precision | Recall | F1-Score |
| SVM | 95.441% | 96% | 95% | 95% |
| AlexNet | 96.661% | 97% | 97% | 97% |

## 7    Discussion and Critique

Kang, Choo, and Kang mention that the One-Versus-One classification technique is at times preferred, as it outperforms other approaches in many cases [10]. The One-Versus-One classifier constructs one classifier per pair of classes. At prediction time, the class that receives the most votes is selected [15]. The SVM model will categorize the different points based on the attributes each class exhibits or some mapping function, depending on the size of the kernel used [7].

The SVM model draws a decision boundary between the classes, which acts as the border for new data points coming into the model. From Table 3, the SVM algorithm performance is noted using the accuracy, precision, recall, and f1-score. A model accuracy of 94.441% was produced, which illustrates the algorithm's ability to make distinctions between the two classes correctly. The precision score from Table 3 and Table 1 illustrate the algorithm's ability to predict correct positive observations to total observations with a score of 96%.

The SVM ROC curve performs well, recording an area under the curve of 98.64%, which further demonstrates the algorithm's ability to make classifications between the two classes correctly. The SVM algorithm is well suited for the domain as it manages to effectively and sufficiently classify the stroke class from the gameplay class. While the SVM algorithm remains a stable solution for the problem, the relatively newer CNN architecture outperforms the SVM algorithm marginally. It would appear to be better suited for the domain.

**Fig. 4.** An example of a gameplay image falsely predicted as stroke, where part of the batsman is in the frame, image taken from a YouTube source [2].

From Table 3, the AlexNet architecture recorded an exceptional model accuracy of 96.661%, which outperforms the SVM algorithms accuracy. Additionally, precision, recall, and f1-scores of 97% further contribute to its success. The main reason for the AlexNet architecture success over SVM is its ability to predict fewer false positives, supported by the AlexNet precision score of 97% opposed to SVM's 95%. For both models, the gameplay class exhibits a higher number of false positives, upon inspection of these images it is seen that there is a part of the batsman performing a stroke, hence the resultant cause for the miss-classifications, as seen in Fig. 4. From Table 2, the AlexNet confusion matrix had fewer false positives than that of the SVM algorithm, as it had six fewer false positives for the gameplay class and two fewer false positives for the stroke class. Even though the number of false positives differs slightly between the two pipelines, it is one of the main reasons that the AlexNet architecture was able to outperform the SVM algorithm.

## 8  Conclusion

Recent works have paid special attention to automated analysis within the cricketing domain. However, identifying significant scenes in a video depicting sporting events efficiently remains a challenge. This research aims to create a model that can recognize scenes that contain strokes performed by a batsman in a cricket game. The AlexNet architecture proposed in this research manages to make improvements to current works presented by Kolekar and Palaniappan. The architecture produces recall and precision scores of 97% each. In addition to understanding the interest associated with performing strokes, the model proposed in this research illustrates the potential benefits that may be derived.

The two pipelines implemented was the SVM pipeline and the AlexNet pipeline. Both pipelines produced promising results. The AlexNet architecture produces a 96.661% model accuracy, which outperforms the SVM algorithm that produces a 95.441% model accuracy. The AlexNet architecture producing fewer false positive observations further highlights its ability to outperform the SVM algorithm and serve as a solution to the research problem.

By achieving scene recognition, the research was able to automate the filtering of important events. The filtering of significant events is also being investigated by Kolekar et al., which emphasizes the importance of pruning large volumes of data. Furthermore, contributions toward detecting when a significant event occurs were made as it is established that strokes performed impacts the result of a cricket match. Having noted the success of using CNN's, future works would further analyze CNN architectures such as VGG16, GoogleLeNet, and more. In the research, only two classes were identified for classification, by addressing additional classes, the model may cater to a broader range of different scenes within the domain.

## References

1. Articles, C.T.: Cricket basics explanation. http://www.chicagotribune.com/chi-cricket-basics-explanation-gfx-20150215-htmlstory.html
2. cricket.com.au: second test: Australia v england, day three. YouTube, December 2017. https://www.youtube.com/watch?v=7jElHzWzqAk&t=5s
3. Fawcett, T.: An introduction to ROC analysis. Pattern Recogn. Lett. **27**(8), 861–874 (2006)
4. Fei-Fei, L., Li, L.J.: What, where and who? Telling the story of an image by activity classification, scene recognition and object categorization. In: Cipolla, R., Battiato, S., Farinella, G.M. (eds.) Computer Vision. Studies in Computational Intelligence, vol. 285 pp. 157–171. Springer, Heidelberg (2010). https://doi.org/10.1007/978-3-642-12848-6_6
5. Goodfellow, I., Bengio, Y., Courville, A.: Deep Learning. MIT Press, Cambridge (2016)
6. Goutte, C., Gaussier, E.: A probabilistic interpretation of precision, recall and $F$-score, with implication for evaluation. In: Losada, D.E., Fernández-Luna, J.M. (eds.) ECIR 2005. LNCS, vol. 3408, pp. 345–359. Springer, Heidelberg (2005). https://doi.org/10.1007/978-3-540-31865-1_25
7. Gunn, S.R., et al.: Support vector machines for classification and regression. ISIS Tech. Rep. **14**(1), 5–16 (1998)
8. Hsu, C.W., Chang, C.C., Lin, C.J., et al.: A practical guide to support vector classification (2003)
9. Huggi, S.K., Nandyal, S.: Detecting events in cricket videos using RF classifier. Int. J. Adv. Res. Found. **3**, 1–5 (2016)

10. Kang, S., Cho, S., Kang, P.: Constructing a multi-class classifier using one-against-one approach with different binary classifiers. Neurocomputing **149**, 677–682 (2015)

11. Kapela, R., McGuinness, K., O'Connor, N.E.: Real-time field sports scene classification using colour and frequency space decompositions. J. Real-Time Image Process. **13**(4), 725–737 (2017). https://doi.org/10.1007/s11554-014-0437-7

12. Knight, J.: Cricket for Dummies. Wiley, Hoboken (2013)

13. Kolekar, M.H., Palaniappan, K., Sengupta, S.: Semantic event detection and classification in cricket video sequence. In: Sixth Indian Conference on Computer Vision, Graphics & Image Processing, ICVGIP 2008, pp. 382–389. IEEE (2008)

14. Kondratenko, Y.P.: Robotics, automation and information systems: future perspectives and correlation with culture, sport and life science. In: Gil-Lafuente, A.M., Zopounidis, C. (eds.) Decision Making and Knowledge Decision Support Systems. LNEMS, vol. 675, pp. 43–55. Springer, Cham (2015). https://doi.org/10.1007/978-3-319-03907-7_6

15. Pedregosa, F., et al.: Scikit-learn: Machine learning in Python. Journal of Machine Learning Research **12**, 2825–2830 (2011)

16. Saha, S.: A comprehensive guide to convolutional neural networks. Technical report, Medium (2018)

17. Suard, F., Rakotomamonjy, A., Bensrhair, A., Broggi, A.: Pedestrian detection using infrared images and histograms of oriented gradients. In: 2006 IEEE Intelligent Vehicles Symposium, pp. 206–212. IEEE (2006)

18. Van De Sande, K., Gevers, T., Snoek, C.: Evaluating color descriptors for object and scene recognition. IEEE Trans. Pattern Anal. Mach. Intell. **32**(9), 1582–1596 (2009)

19. Wencheng, C., Xiaopeng, G., Hong, S., Limin, Z.: Offline Chinese signature verification based on AlexNet. In: Sun, G., Liu, S. (eds.) ADHIP 2017. LNICST, vol. 219, pp. 33–37. Springer, Cham (2018). https://doi.org/10.1007/978-3-319-73317-3_5

20. Zhou, W., Gao, S., Zhang, L., Lou, X.: Histogram of oriented gradients feature extraction from raw Bayer pattern images. IEEE Trans. Circuits Syst. II Express Briefs **67**, 946–950 (2020)

# Biologically Inspired Modelling of Flowers

Krzysztof Najda and Łukasz Dąbała$^{(\boxtimes)}$ (iD)

Warsaw University of Technology, pl. Politechniki 1, 00-661 Warsaw, Poland
l.dabala@ii.pw.edu.pl

**Abstract.** In this work we present a tool for three-dimensional procedural modelling of flowers based on botanical information. The main biological input is floral formulas. It can represent the structure of a flower with numbers, letters and symbols. Based on this notation, a user-friendly interface was created. Artists need to draw several curves and provide floral formula to create a full 3d model. It speeds up the recreation of biologically-correct models and thanks to its implementation in Maya, each flower can be easily customized after creating initial model.

**Keywords:** Geometry · Procedural generation · Biologically correct models

## 1 Introduction

Flowers are very interesting objects to create. Their variety comes from great number of components, that every artist needs to recreate - starting from petals and ending with stamen. What is more, depending on the flower, the number of components can be different, as well as their relative position or size. This is a great challenge for rapid development of models - how to achieve simplicity and plausibility and at the same time to be consistent with biology.

Existing methods can be divided into two main categories: focused on the effect and correctness with nature. The first one can be represented by the work [6], where authors proposed interface for modeling with predefined components. That simplifies the whole process, but limits the number of flowers that can be created. An example of the second category are the L-systems [10], where authors used symmetrical properties of flowers. They defined mathematical rules to create plants, but here user needs to know a lot about the biology. Authors in [3] simplified the process and used the floral diagrams and inflorescences to create the whole plants with the flowers. They managed to create the application with simple interface and at the same time it allowed to use biological information.

In our work, we tried to simplify the process of modelling flowers a little bit more and try to use different kinds of biological information - floral formulas instead of diagrams. They should be correspondingly similar, but rewriting the formula from the book should be a much quicker way to reconstruct desirable objects. Our system is implemented as a plugin to the commercial software, which creates the possibility to open it to public and spread it.

© Springer Nature Switzerland AG 2020
L. J. Chmielewski et al. (Eds.): ICCVG 2020, LNCS 12334, pp. 110–120, 2020.
https://doi.org/10.1007/978-3-030-59006-2_10

# 2    Background and Previous Work

## 2.1    Floral Equations

Various biologist tend to find a way to describe the structure of the flowers. One of them is floral formula, which consist of letters, symbols and numbers that characterize each species. This notation has been present for a long time in the literature - it was developed in 19th century [9].

The floral formula gives the information about the number of elements in the flower, how elements are fused and how they are aligned. It uses several symbols to describe each of possible part of the flower: perianth, calyx, corolla, stamen and pistil. Symbols can also give information about symmetry of the flower and its type.

Because of the fact, that the format of the formula can differ between authors [1], we also propose one, which will be described in following sections.

## 2.2    Previously Created Systems

L-systems are the most common systems used for modelling flowers [10]. Such systems were developed over the years and researchers added more and more functionality such as: interaction with environment [7]. This kind of systems can produce nice looking plants, but the problem lies in the definition of the rules for the growth. What is more, such systems are rather used to create branching systems than flowers themselves.

Another type of methods are sketch-based systems. In these type of methods, an user needs to provide several curves to create the whole model. Usually, there is some additional information involved. One of such systems is the one mentioned before: one created by Ijiri et al. [3], where they made the whole system for flowers creation using floral diagrams. Researchers, also used sketches to create trees [8]. Such systems allow users to create botanically correct models, but without proper implementation it can be problematic to modify models.

People were trying to use various techniques to obtain the data about the plants and create semi-automatic systems for their recreation. Basic data involves photos and video. In the paper by Tan et al. [11], they created a method, that allowed user to generate a 3d model of a tree based on a single photo and a couple of user strokes. Sometimes, more focus was put into specific parts of the flower like petals. In the work by Zhang [12], they achieved a realistic petals from a 3d point cloud coming from a single view. To obtain such result, first they segmented each petal and used special scale-invariant morphable petal model to reconstruct them well. They also involved botanical information to correct occluded regions and spatial relations.

Some of the researchers went a step further in the reconstruction process. In work by Ihiri et al. [4], they used computed tomography as the data source and they specially designed an active contour model for flowers. It allowed them to reconstruct realistic flowers with thin petals.

Image based as well as computed tomography based systems have their own pros and cons. The first ones have difficulties in obtaining information about the occluded regions. In the flowers, there are many places, that will not be visible for a camera, so without additional information the reconstruction will not be correct.

The best results can be produced by computed tomography because we have all needed data for the reconstruction. However it needs special equipment which is not available for the common user.

## 3    Created Method

### 3.1    Modeling Flower

In order to generate 3d mesh of flower we need to find a way to model its appearance. We decided to utilize commonly used flower morphology. According to it, the flower is divided into 5 distinct parts (Fig. 1). We modelled each part separately, which helps to speed up and simplify the recreation process.

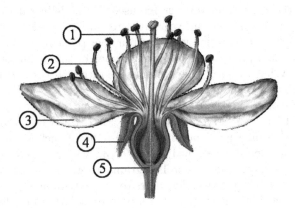

**Fig. 1.** Flower morphology: 1 - carpel, 2 - stamen, 3 - petal, 4 - sepal, 5 - receptacle; source: [5]

Receptacle is a base for all other elements. Carpel is always positioned in the flower's center while other elements are placed on receptacle's edge. Also petals, sepals and stamens are always placed on single or multiple rings around flower's center. Each ring might have different properties - for example petals on different rings might vary in shape, number, color etc. while petals on the same ring are completely uniform.

### 3.2    Floral Equations

As it was described previously, floral equations provide some data about flower. In our project we used the following grammar for this notation (EBNF):

```
floral_equation  =    symmetry calicle stamens carpels receptacle
symmetry         =    '*'  |  '|'  |  '$'
calicle          =    tepals | sepals petals
tepals           =    'P' num
sepals           =    'K' num
petals           =    'C' num
stamens          =    'A' num
carpels          =    'G' num
receptacle       =    ', s'  |  ', i'
num              =    num2 {'+' num2}
num2             =    num3 | '(' num3 ')'
num3             =    INT  |  'n'  |  INT '-' INT  |  INT '-' 'n'
```

Although reading grammar is necessary to understand this notation fully it may prove cumbersome without any examples. And so let's take a look at a following equation: "* P3+3 A3+3 G(3), s". Reading from left to right we can see that this flower has radial symmetry thus we put '*' at the beginning. Reading further we can see a lot of numbers and letters - it's important to keep in mind that numbers indicate number of elements on the ring, while '+' indicates additional ring. Therefore, we see that this flower has two tepal rings which consist of 3 elements ('P3+3'). Similarly there are two further rings consisting of 3 stamens each ('A3+3'). The next one is carpel ('G(3)') which as we can see is triple conjoined. The last information (',s') says that this flower has superior receptacle which means that ovary is surrounded by it.

As it can be seen, floral equations are quite informative and compact at the same time. To take advantage of it we created our own parser that is able to parse through this notation and output parameters in orderly fashion as data structure. This data is later used as a base for further operations which are presented in further sections.

### 3.3   Floral Curves

Unfortunately floral equations do not provide all data required to generate flower's model in 3D, namely:

1. Stamen type
2. Angles
3. Color
4. Shape of each element

The first three can be easily provided by the user because these are numeric values. During implementation of the interface, they can be presented as series of spinboxes or a color picker. The fourth one, however, which is shape of each element, is much more complex. Currently there is no biological information (other than sketc.hes and photos) that would be able to define flower's shape precisely. Therefore we decided to create our own solution to this problem which we named "floral curves".

Although the name may be intimidating, the concept behind it is really simple. Its task is to provide program with array of points that are evenly spaced along a path. These points are interpolated by using knot points which are provided by the user (Fig. 2).

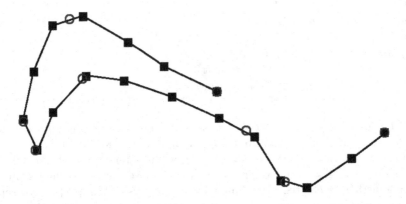

**Fig. 2.** The example of a floral curve: blue circles - knot points, black squares - interpolated points (Color figure online)

From the plethora of interpolation methods we decided to use Hermite interpolation which is a variant of cubic interpolation. It requires only a set of knot points and offers two parameters (tension and bias) that may be manipulated to change a shape of the curve.

There are many other types of potential solutions of the curve representation like Bezier curve, but we wanted to make the interface as simple as possible. It's getting more and more complicated using Bezier curves because it requires from the user to attach additional control points to knot points. We concluded that this level of precision is not necessary and it only makes defining shapes more complex with no apparent benefits.

We elaborate on floral curve significance for each of flower's element in further sections.

### 3.4   Creating Meshes

The central part of our solution is mesh generator. It uses abstract flower model provided by the user (using floral equations, floral curves and other simple data) to create vertices, edges and polygons. In order to accomplish this task we use simple operations such as translation, rotation and scaling, but also some more complex ones, namely: extruding and morphing.

In the context of this paper morphing is a method that takes two cross-sections (represented by points array) with the same number of points and

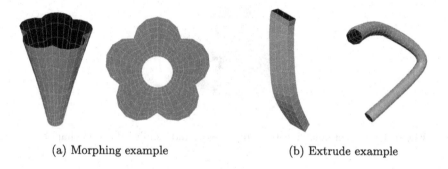

(a) Morphing example                    (b) Extrude example

**Fig. 3.** Main operations used during creation of a model

interpolation factor (ranging from 0 to 1). This method simply uses linear interpolation on each point to create third cross-section. The example is shown on Fig. 3a.

Extruding, on the other hand takes one cross-section and path (in our case floral curve) to create mesh. It's a bit more complex than morphing and operates in following steps:

1. Interpolate given number of points along floral curve
2. Compute vectors between neighboring points
3. Rotate given cross-section along these vectors
4. Move rotated cross-section so that the center of cross-section is placed on path's point
5. Repeat steps 3–5 for each point on floral curve

Example is shown on Fig. 3b.

In our project we don't use just any type of cross-sections. There are four distinct types present, which are as follows:

1. Circle
2. Rectangle
3. "Rosette"
4. "Smile"

Circle and rectangle are too straightforward to be explained and example of "rosette" and "smile" are shown on Fig. 4. Rosette consists of a number of circles merged together, while "smile" is basically curved rectangle. Certain types of cross-sections are used to generate different elements of flower.

### 3.5  Creating Mesh of Each Element

Each of different elements is generated separately and then placed in their respective spot on flower.

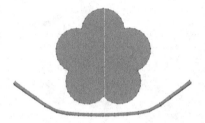

**Fig. 4.** Cross-sections: "rosette" (top shape) and "smile" (bottom shape)

**Carpel Generation.** Carpel is the first object that is being generated. It requires longitudinal section and number of conjoined carpels (which determines how many indentations are visible). Based on that generator is able to create cross-sections along the carpel height which are then joined together to form mesh. These cross-sections are created using morphing of circle and rosette. In the most narrow sections of the carpel, cross-sections resemble circle and the wider the carpel the more its cross-section is similar to rosette.

**Receptacle Generation.** Receptacle is generated using the same algorithm as carpel. The only difference is that it modifies given longitudinal section, cutting it at the height of ovary. That way receptacle is always of acceptable size. Also morphing here is reverted - wider sections morph towards circle, narrow towards rosette.

**Stamen Generation.** Stamen is divided into two separate parts: filament and head. Filament is a circle cross-section which is extruded along given floral curve. It is also getting slightly narrower as it approaches the head of the stamen. On the other hand, stamen's head consists of a given number of ellipsoids put together. Head is then put on top of filament to create full stamen. To avoid feeling repetitive the stamen's floral curve is a bit distorted. Thanks to that each stamen looks ever so slightly different even if they have been generated with the same data.

**Sepal, Petal and Tepal Generation.** These elements have by far the most complex generation algorithm (they use the same one), It consists of three different floral curves and uses morphing and extruding at the same time. These curves define petal, sepal or tepal shape as you can see an example Fig. 5. We accomplish this shape in similar fashion to creating carpel. The difference is that in this case we use rectangle and "smile" to morph consecutive cross-sections instead of circle and rosette. The wider part of calicle the thicker and more curved it is. Also these cross-sections are being extruded along one of floral curves instead of straight line.

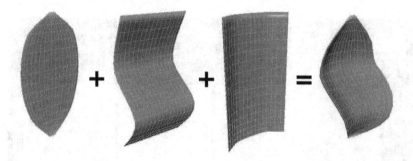

**Fig. 5.** Calicle generation

## 3.6  Algorithm Summary

The whole algorithm from floral equation to full 3d mesh works as follows:

1. Parse floral formula
2. Set biological information not included in floral formula (such as angles of each ring)
3. Set floral curves for each element
4. Set mesh resolution to desirable level
5. Run mesh generator.

# 4  Results

Thanks to our solution we are able to recreate a wide variety of flowers in a matter of minutes. We will demonstrate process of flower generation using an example of apple blossom (Fig. 6)

**Fig. 6.** Apple blossom generated with our solution

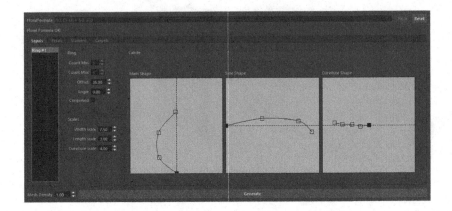

**Fig. 7.** User interface

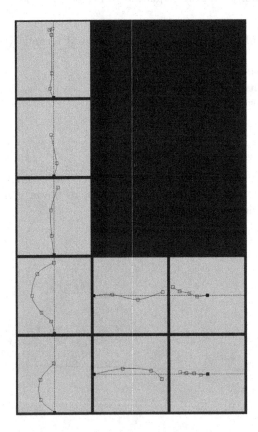

**Fig. 8.** Floral curves used to generate apple blossom. Their meaning from top to bottom: carpel, stamen #1, stamen #2, petal, sepal. In case of petal and sepal we use three different floral curves, their meaning from left to right: width, side, front

**Fig. 9.** Examples of models created with the proposed method

It has been generated using the following floral formula: "*K5 C5 An + 6-8 G(2)". By reading this equation we can see that this flower has radial symmetry ('*'). It consists of 5 free sepals ('K5') and petals ('C5'). There are two rings of stamens (An + 6-8) one of which has ¿12 stamens on it ('n') and number of elements on the second one ranges from 6 to eight ('6–8'). We can also see that carpel is double conjoined ('G(2)' - the brackets define fusion of elements). That way we described concisely what kind and number of elements are present in our flower.

However, to define its shape we need to get into the next major step - drawing floral curves. The exact floral curves used in this instance are shown on Fig. 8. Apart from tweaking some other values such as scales, angles and offsets (most of which are just set by trial and error) it is all data which is needed to generate any flower. The UI of our Maya plugin can be seen on Fig. 7.

Some other pictures of flowers generated with our solution can be seen on Fig. 9.

## 5   Conclusions and Future Work

We managed to create a plugin for commercial 3d software that allows user to create flowers easily. Generated models use biological information to make the final creation as close as possible to reality. Due to the fact that each element is defined separately the user has control over the process of generation.

During creation of our tool for the artist we came to the conclusion, that one thing can be added to make the solution complete. That is automatic texturing based on the photo or video. Due to the specificity of the flower structure, such a system should involve operations like semantic segmentation and lighting decomposition. This is caused by many occlusions caused by petals and ambiguities in flower appearance. We recommend using a neural network like Mask R-CNN [2] to make it possible.

# References

1. De Craene, L.: Floral Diagrams: An Aid to Understanding Flower Morphology and Evolution. Cambridge University Press (2010). https://books.google.pl/books?id=24p-LgWPA50C
2. He, K., Gkioxari, G., Dollár, P., Girshick, R.B.: Mask R-CNN. In: 2017 IEEE International Conference on Computer Vision (ICCV), pp. 2980–2988 (2017)
3. Ijiri, T., Owada, S., Okabe, M., Igarashi, T.: Floral diagrams and inflorescences: interactive flower modeling using botanical structural constraints. In: ACM SIGGRAPH 2005 Papers, SIGGRAPH 2005, New York, NY, USA, pp. 720–726. Association for Computing Machinery (2005). https://doi.org/10.1145/1186822.1073253
4. Ijiri, T., Yoshizawa, S., Yokota, H., Igarashi, T.: Flower modeling via X-ray computed tomography. ACM Trans. Graph. 33(4) (2014). https://doi.org/10.1145/2601097.2601124
5. Holeczek, J., Pawłowska, J.: Puls życia 5. Nowa Era, January 2018
6. Lintermann, B., Deussen, O.: Interactive modeling of plants. IEEE Comput. Graph. Appl. 19(1), 56–65 (1999). https://doi.org/10.1109/38.736469
7. Mech, R., Prusinkiewicz, P.: Visual models of plants interacting with their environment. In: Computer Graphics (SIGGRAPH 1996), September 2000
8. Okabe, M., Owada, S., Igarashi, T.: Interactive design of botanical trees using freehand sketches and example-based editing. Comput. Graph. Forum 24(3), 487–496 (2005). https://doi.org/10.1111/j.1467-8659.2005.00874.x
9. Prenner, G., Bateman, R., Rudall, P.: Floral formulae updated for routine inclusion in formal taxonomic descriptions. Taxon 59, 241–250 (2010). https://doi.org/10.2307/27757066
10. Prusinkiewicz, P., Lindenmayer, A.: The Algorithmic Beauty of Plants. Springer, Berlin (1996)
11. Tan, P., Fang, T., Xiao, J., Zhao, P., Quan, L.: Single image tree modeling. In: ACM SIGGRAPH Asia 2008 Papers, SIGGRAPH Asia 2008. Association for Computing Machinery, New York (2008). https://doi.org/10.1145/1457515.1409061
12. Zhang, C., Ye, M., Fu, B., Yang, R.: Data-driven flower petal modeling with botany priors. In: 2014 IEEE Conference on Computer Vision and Pattern Recognition, pp. 636–643 (2014)

# Tuberculosis Abnormality Detection in Chest X-Rays: A Deep Learning Approach

Mustapha Oloko-Oba and Serestina Viriri[(✉)] [iD]

School of Mathematics, Statistics and Computer Sciences,
University of KwaZulu-Natal, Durban, South Africa
219098624@stu.ukzn.ac.za, viriris@ukzn.ac.za

**Abstract.** Tuberculosis has claimed many lives, especially in developing countries. While treatment is possible, it requires an accurate diagnosis to detect the presence of tuberculosis. Several screening techniques exist and the most reliable is the chest X-ray but the necessary radiological expertise for accurately interpreting the chest X-ray images is lacking. The task of manual examination of large chest X-ray images by radiologists is time-consuming and could result in misdiagnosis as a result of a lack of expertise. Hence, a computer-aided diagnosis could perform this task quickly, accurately and drastically improve the ability to diagnose correctly and ultimately treat the disease earlier. As a result of the complexity that surrounds the manual diagnosis of chest X-ray, we propose a model that employs the use of learning algorithm (Convolutional Neural Network) to effectively learn the features associated with tuberculosis and make corresponding accurate predictions. Our model achieved 87.8% accuracy in classifying chest X-ray into abnormal and normal classes and validated against the ground-truth. Our model expresses a promising pathway in solving the diagnosis issue in early detection of tuberculosis manifestation and, hope for the radiologists and medical healthcare facilities in the developing countries.

**Keywords:** Tuberculosis · Chest X-ray · Classification · Deep learning · Pulmonary · Convolutional neural network

## 1 Introduction

Tuberculosis, known as TB, is regarded as a significant health challenge across the world but more prevalent in developing countries [1–3]. TB is caused by the Mycobacterium tuberculosis bacteria that generally affect the lungs (pulmonary), and various parts of the body (extrapulmonary) [4]. Several TB patients lose their lives yearly as a result of diagnosis error and lack of treatments [1,5].

Pulmonary Tuberculosis is a transmissible infection mostly visible around the collarbone. It is contractible via the air when people who have an active

© Springer Nature Switzerland AG 2020
L. J. Chmielewski et al. (Eds.): ICCVG 2020, LNCS 12334, pp. 121–132, 2020.
https://doi.org/10.1007/978-3-030-59006-2_11

tuberculosis infection sneeze, cough, or transmit their saliva through the air or other means [6]. In other words, persistence and closeness of contact are major factors of the risk of contracting TB which make individuals inhabiting in the same household at higher risk of contracting TB than casual contacts. Lack of diagnosis and treatment of affected individuals will as well increase the rate of transmission.

Tuberculosis disease is certainly treatable if identified/diagnosed early for appropriate treatment. Diverse test procedures may be employed to confirm a diagnosis of suspected pulmonary tuberculosis. These test techniques include a Computed Tomography (CT) scan, Chest X-ray (CXR), Magnetic Resonance Imaging (MRI) scan or ultrasound scan of the affected part of the body. Among the various screening techniques in existence, chest X-Ray is renowned for evaluating the lungs, heart and chest wall to diagnose symptoms as reported by [7] and recommend their application for screening patients in order to exclude the usual instance of a costly test.

World Health Organization affirms that Tuberculosis is, in fact, one of the topmost causes of death across the world most especially in developing countries. As reported in 2016, 1.8 million deaths were recorded resulting from about 10.4 million cases of people affected with tuberculosis. As much as tuberculosis can be detected on chest X-ray, tuberculosis prevalent regions usually suffer the expertise of radiologists required to accurately diagnose and or interpret the X-ray results [2].

In this paper, we propose a model that will accurately improve the quality and timely diagnosis of chest X-rays for the manifestation of tuberculosis. The proposed model will increase diagnosis efficiency, enhance performance and eventually minimize cost as opposed to the process of manually examining the chest X-ray scan which is costly, time-consuming, and prone to errors due to lack of professional radiologist and volume of the chest X-rays.

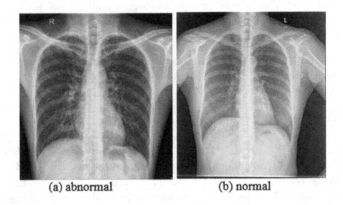

(a) abnormal             (b) normal

**Fig. 1.** Sample of the normal and abnormal chest X-rays.

## 2    Related Work

The strengths and possibilities offered by machine learning has provided a boost to computer vision especially for the diagnosis and screening of various health diseases and conditions. In spite of the global application of machine learning techniques in the biomedical domain, chest X-ray is still a very important and renowned tool [7] among others for evaluating pulmonary diseases that require rapt attention. There is always a need to improve the existing methods and proposing new techniques for stability, global growth, and better performance.

A handful of investigation has been done by [8–10] in assessing the ability of the existing computer-aided detection CAD systems in biomedical domain to diagnose pulmonary nodules. Notable improvement of 0.986 area under curve (AUC) was reported in [8] with the use of a computer-aided design system as opposed to 0.924 AUC without CADs. Similarly, 0.923 AUC improvement with CADs was observed by [9] over 0.896 AUC without CADs. The results of their assessment ultimately show the significant impact of CADs in assisting radiologists to improve the diagnosis from chest X-rays.

An experiment for tuberculosis screening was presented in [24] using Alexnet and VGGNet architectures for the classification of CXR into positive and negative classes. The analysis carried out on the Montgomery and Shenzhen CXR datasets reported that VGGNet outperformed Alexnet as a result of a deeper network of VGGNet. The performance accuracy of 80.4% was obtained for Alexnet while VGGNet reached 81.6% accuracy. The authors concluded that improved performance accuracy is possible by increasing the dataset size used for the experiment.

Another article that deals with detecting the presence of tuberculosis early is presented in [14] which applied a median filter, histogram equalization and homomorphic filter to preprocess the input image before segmentation is applied utilizing the active contour and finally arriving at the classification using the mean values. Although, the research does not report any accuracy attained but affirm the impact and contribution of computer-aided diagnosis as an assisting tool to guide radiologists and doctors in reaching an accurate and timely diagnosis decision with respect to tuberculosis detection.

The study carried out in [11] presented proposals that will improve feature extractors in detecting diseases using pre-trained CNN. This research is famous for combining multiple instance learning algorithms with pre-trained CNN, assessment of classifiers trained on features extracted, and the comparison of performance analysis of existing models for extracting features from the radiograph dataset and achieved an accuracy of 82.6%.

An experiment was done [15] to compare the performance of three different methods in detecting tuberculosis in the lungs. The obtained results show the K-nearest neighbor (KNN) classifier with the maximum accuracy of 80%, Simple linear regression 79%, and the Sequential minimal optimizer 75% accuracy.

An approach to discover tuberculosis from a radiograph image where the X-ray machine is placed behind the patients (posteroanterior) was presented in [12]. This study employs a graph cut segmentation approach for extracting the lung region on the chest X-ray and then computes sets of features such as (edge, texture, and shape) that allow classification into abnormal and normal classes respectively.

A CNN model that involved classification of different manifestations of tuberculosis was presented in [19]. This work looked at unbalanced and less categorized X-ray scans and incorporated cross-validation with sample shuffling in training the model and reported to have obtained 85.6% accuracy in classifying the various manifestation of tuberculosis using the Peruvian datasets composed of a total of 4701 samples with about 4248 samples labeled as abnormal containing six manifestation of tuberculosis and 453 samples labeled as normal. The authors affirm that their model surpasses previous classification models and is promising in tuberculosis diagnosis.

The research work presented in [13] is one of the first to employ deep learning techniques on medical images. The work was based on popular Alexnet architecture and transfer learning for screening the system performance on different datasets. The cross dataset performance analysis carried out shows the system accuracy of 67.4% on the Montgomery dataset and 83.7% accuracy on the Shenzhen dataset.

In [25], the authors participated in the 2019 ImageCLEF challenge and presented a deep learner model (LungNet) that focus on automatic analysis of tuberculosis from computer tomography CT scans. The CT scans employed is firstly decompressed and the slices are extracted having 512 images for the X and Y dimensions and between 40–250 images for the Z dimension. Filters were then introduced on the slices to eliminate the slices that do not contain valuable information required for classifying the samples. The proposed LungNet along with ResNet-50 architecture was employed as a deep learner whose outputs are regarded as the preliminary results. The deep learner model was trained on 70(%) and 50(%) training sets and achieved AUC performance of 63(%) and 65(%) for the ImageCLEF CT report and severity scoring task respectively. Although, these performances were not the best presented in the challenge but the authors believed if subjected to advanced pre-processing techniques such as data augmentation, and masking could provide a better performance.

CheXNeXt is an algorithm developed by the authors in [26], for identification of 14 various pathologies in chest X-rays. The algorithm which employed convolutional neural network approach was validated on the NIH dataset and compared the result with interpretation of 9 professional radiologists. The results show that CheXNeXt achieved equivalent performance with radiologists in 10 different pathologists, best performance on 1 pathology (atelectasis) attaining 0.862(%) AUC and underperformed in 3 pathologists. The algorithm took less than 2 min to identify the various pathologists while it took the radiologist about 240 min.

The study in [27] is conducted to assess the detection accuracy of qXR, a computer aided diagnosis software based on convolutional neural network. The authors utilized microbiologically established lung tuberculosis images as the standard for reference and made use of kappa coefficient along with confidence interval as the statistical tools to analyse the data and examine the inter-rater reliability of radiologist in detecting certain lung abnormalities. The study also used radiologist interpretation as standard to validate the detection accuracy of the qXR in terms of generating ROC curves and calculating AUC. The qXR system achieved 0.81(%) AUC for detection of lung tuberculosis, 71(%) sensitivity and 80(%) specificity.

As with existing models, most of which basically show the performance accuracy of their models without a view of the predicted sample. We, however in this work present a model evaluated on the Shenzhen datasets to provide an improved performance accuracy thereby showing predictions of the model validated on the groundtruth data. The groundtruth data will ultimately assist us to further develop a tuberculosis diagnosis system to be deployed to health facilities in the developing countries to improve quality and timely diagnosis.

## 3    Methods and Techniques

### 3.1    Datasets

The dataset used in training our model is the Shenzhen tuberculosis dataset. This dataset is specific to tuberculosis and is publicly available for the purpose of research. The Shenzhen dataset is made up of 336 abnormal samples labeled as "1" and 326 normal samples labeled as "0". All samples are of the size 3000 by 3000 pixels saved as portable network graphic (png) file format as shown in Fig. 1. This dataset is accompanied by clinical readings that gives details about each of the samples with respect to sex, age, and diagnosis. The dataset can be accessed at https://lhncbc.nlm.nih.gov/publication/pub9931.

### 3.2    Convolutional Neural Networks

A broadly used deep learning algorithm for object classification, detection, and video analyzing in computer vision is the Convolutional Neural Network otherwise known as ConvNet or CNN. ConvNet is able to distinguish one object

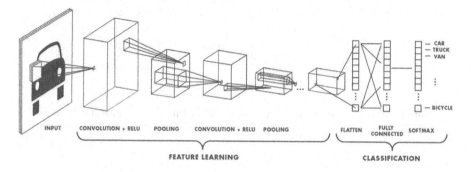

**Fig. 2.** CNN architecture extracted from [18].

from another by assigning important learnable biases and weights to various images [19]. ConvNets which are feed-forward neural networks are famous for performing better than other classification networks as a result of taking advantage of the inherent properties of the images. The required preprocessing in a convolutional neural network is less compared to other classification neural networks. The ConvNet is made of an input object, features learning layers such as (conv2d, pooling, normalization) with ReLu activation function and finally the classification layers such as (flatten, fully connected, dense) with softmax or sigmoid activation functions as shown in Fig. 2.

### 3.3  Preprocessing

ConvNet models performs better with large dataset [17], hence we applied augmentation [16] as a preprocessing methods to increase the size of data for better performance. The augmentation types we applied to enlarge the size of our data are: flip_left_right with a probability = 0.5, zoom_random = 0.5, with an area = 0.8, and flip_top_bottom = 0.5, rotate left_right = 5. With the use of data augmentation, we were able to enlarge our dataset to 10,000 samples.

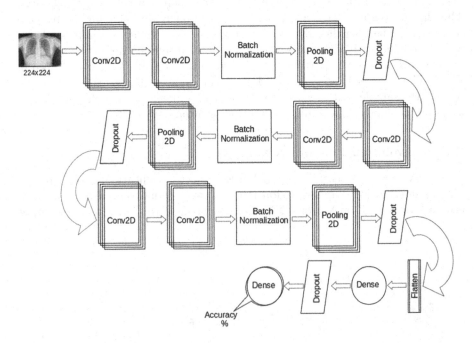

**Fig. 3.** Architecture of our 6-layer network for tuberculosis classification.

## 3.4   Model Architecture

Our ConvNet model is a sequential model that is made up of 6 convolutional layers, batch normalization, dense layer sas shown in Fig. 3. Each layer in the model use the Rectified Linear Unit (ReLu) activation function except the second dense layer which serves as the output layer and used the Softmax activation function written mathematically as:

$$\beta(v)_i = \frac{e^{v_i}}{\sum_{n=1}^{n} e_n^v}$$

where $v$ mean the input vector to the output layer $i$ that is depicted from the exponential element.

The input shape is $224 \times 224 \times 3$ which is fed to the 1st convolutional layer using a $32 \times 3 \times 3$ filters, the output from the first layer is pass to the 2nd convolutional layer which also makes use of a $32 \times 3 \times 3$ filters. This is followed by the next two convolutional layers which employ a $64 \times 3 \times 3$ filters and the output from here is fed to the final 2 convolutional layers with a $128 \times 3 \times 3$ filters respectively. The output from the last convolutional layer is normalized, flatten and pass to the dense layer with 4096 neurons. A regularization technique dropout of $(0.5)$ was added to control overfitting before passing to the last dense layer which is the output layer that uses softmax activation with 2 neurons.

The activities in the convolutional layer are given as:

$$p_j^i = f\left(\sum_{i \in N_j} p_i^{l-1} * Q_{ij}^l + \alpha_j^l\right)$$

where $p_j^i$ depicts the $j^{th}$ output feature of the $l^{th}$ layer, $f(.)$ is a nonlinear function, $N_j$ is the input map selection, $p_i^{l-1}$ refers to the $i^{th}$ input map of $l-1^{th}$ layer, $Q_{ij}^l$ is the kernel for the input $i$ and output map $Q$ in the $j^{th}$ layer, and $\alpha_j^l$ is the addictive bias associated with the $j^{th}$ map output.

## 3.5   Training

The dataset was split into a 70% training set, 10% validation set and 20% testing set and compile our model using cross-entropy loss function and the Stochastic Gradient Descent (SGD) optimizer along with varieties of hyper-parameters. The initial learning rate was set to $= 0.01$, momentum $= 0.9$ with regularization L2 $= 0.0004$ as a starting point to control overfitting and training loss and we measure our metrics by the accuracy obtained from the best experiment. The SGD optimizer is given below as:

$$\beta = \beta - x. \bigtriangledown_\beta J(\beta; n^{(i)}; m^{(i)})$$

where $\bigtriangledown_\beta J$ is the gradient of the loss w.r.t $\beta$, $x$ is the defined learning rate, $\beta$ is the weight vector while $n$ and $m$ are the respective training sample and label.

The model was trained using Keras deep learning framework running on tensorflow backend. The training is done entirely in the cloud on a Telsa GPU provided by Google which is available via google col laboratory notebook platform. Collaboratory is a free cloud Jupyter notebook platform that provides researchers with free GPU up to 12 GB of random access memory (RAM) and 360 GB of hard disc.

## 4    Results

We measure the performance of our model using the accuracy metric which determined the ratio of the correctly predicted samples from the entire predictions made. This metric is given as:

$$Accuracy = \frac{TP + TN}{TP + FP + FN + TN}$$

where

TP: is the case where the model predicts a sample as positive and it is actually positive
TN: is the case where the model predicts a sample as negative and it is actually negative
FP: is the case where the model predicts a sample as positive whereas the actual output is negative
FN: is the case where the model predicts a sample as negative whereas the actual output is positive

The training and validation of our proposed model were performed using the Shenzhen TB specific dataset. The dataset which consists of a total of 662 samples were augmented to 10,000 samples out of which 70% were used for training while 30% was utilized for testing validation. The training samples were shuffled during model training to make certain that both positive and negative samples were included in the training to avoid biases. We performed several experiments and obtained 87.8% performance accuracy.

The initial experiment was performed with 662 samples and obtained 79(%) accuracy whereas after applying some image processing techniques, our model achieved 87.8(%) accuracy. We applied contrast limited adaptive histogram equalization on the chest X-ray samples to enhance the image quality which made features identification and extraction distinctly clearer and obvious to the classification module for accurate classification.

The data augmentation and CLAHE techniques were some factors responsible for obtaining an improved performance. Samples of the enhance images is presented in Fig. 6 which shows obvious clarity compared to the samples in Fig. 1.

The loss and performance accuracy of the model is presented in Fig. 4. In Table 1 is the summary of the existing and proposed CNN models with their accuracy while Fig. 5 shows few predictions of the proposed model.

**Table 1.** Summary of tuberculosis detection models showing the various datasets, assessment measure and the performance accuracy

| Authors ref | Datasets | Evaluation metric | Accuracy (%) |
|---|---|---|---|
| [13] | Montgomery County and Shenzhen | Accuracy | 67.4, 83.7 |
| [11] | Shenzhen and Montgomery County | Accuracy | 84.7, 82.6 |
| [19] | Peruvian partners at "Socios en Salud". (Peru) | Accuracy | 85.6 |
| [24] | Montgomery County and Shenzhen | Accuracy | 80.4, 81.6 |
| [25] | Custom CT scan | Accuracy, AUC | 63, 65 |
| [26] | NIH Chest X-ray14 | AUC | 86.2 |
| [27] | Custom: Chest x-ray at Kasturba hospital Manipal, India | AUC, Sensitivity, Specificity | 81, 71, 80 |
| **Proposed** | **Shenzhen** | **Accuracy** | **87.8** |

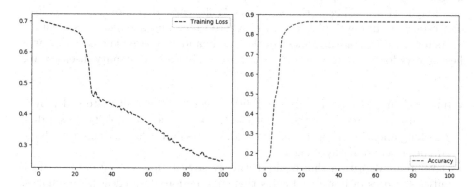

**Fig. 4.** Training loss and accuracy

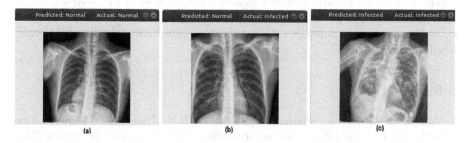

**Fig. 5.** Model predictions. These are few predictions of our model. (a) and (c) are samples that were correctly predicted while (b) is one of the very few misclassified sample that will be improve in future work

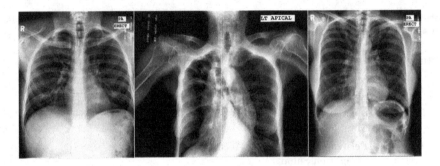

**Fig. 6.** Samples of the enhanced chest X-ray

## 5    Conclusion

We presented a ConvNet model in this paper for improving the diagnosis of pulmonary tuberculosis from chest X-ray images for prompt and effective treatment. The ConvNet model presented here achieved 87.8% classification accuracy and further shows the prediction. Comparing the predicted samples with the ground-truth has expressed an impressive performance which is promising to the thought of providing lasting solutions to the early detection of tuberculosis most especially in developing countries where tuberculosis is most prevalent.

As for the future advancement in the timely and accurate diagnosis of a large chest X-rays for the detection of tuberculosis and other pulmonary diseases. We intend to:

- Improve on our model to reach 100(%) accuracy and ultimately deploy a tuberculosis diagnosis system to health facilities where radiological expertise is lacking and hindering the quality of diagnosis and delay in diagnosing large number of chest X-ray images.
- Develop an algorithm that could distinctly identify foreign objects like rings, buttons, pieces of bone, and coins that may be found on chest x-ray images. This foreign objects can results in misclassification and will impede performance of detection systems. For instance round objects like buttons, bone piece or rings can be mistaken for nodules.
- Deploy a computer-aided detection systems to simultaneously detect different pulmonary diseases.

## References

1. Zumla, A., George, A., Sharma, V., Herbert, R.H.N., Oxley, A., Oliver, M.: The WHO 2014 global tuberculosis report-further to go. Lancet Glob. Health **3**(1), e10–e12 (2015)
2. Lakhani, P., Sundaram, B.: Deep learning at chest radiography: automated classification of pulmonary tuberculosis by using convolutional neural networks. Radiology **284**(2), 574–582 (2017)

3. Sathitratanacheewin, S., Pongpirul, K.: Deep learning for automated classification of tuberculosis-related chest X-ray: dataset specificity limits diagnostic performance generalizability. arXiv preprint arXiv:1811.07985 (2018)

4. Baral, S.C., Karki, D.K., Newell, J.N.: Causes of stigma and discrimination associated with tuberculosis in Nepal: a qualitative study. BMC Public Health **7**(1), 211 (2007)

5. Hooda, R., Sofat, S., Kaur, S., Mittal, A., Meriaudeau, F.: Deep-learning: a potential method for tuberculosis detection using chest radiography. In: 2017 IEEE International Conference on Signal and Image Processing Applications (ICSIPA), pp. 497–502 (2017)

6. Bhatt, M.L.B., Kant, S., Bhaskar, R.: Pulmonary tuberculosis as differential diagnosis of lung cancer. South Asian J. Cancer **1**(1), 36 (2012)

7. World Health Organization: Chest radiography in tuberculosis detection: summary of current WHO recommendations and guidance on programmatic approaches (No. WHO/HTM/TB/2016.20). World Health Organization (2016)

8. Kakeda, S., et al.: Improved detection of lung nodules on chest radiographs using a commercial computer-aided diagnosis system. Am. J. Roentgenol. **182**(2), 505–510 (2004)

9. Sakai, S., et al.: Computer-aided nodule detection on digital chest radiography: validation test on consecutive T1 cases of resectable lung cancer. J. Digit. Imaging **19**(4), 376–382 (2006). https://doi.org/10.1007/s10278-006-0626-4

10. Shiraishi, J., Abe, H., Li, F., Engelmann, R., MacMahon, H., Doi, K.: Computer-aided diagnosis for the detection and classification of lung cancers on chest radiographs: ROC analysis of radiologists' performance. Acad. Radiol. **13**(8), 995–1003 (2006)

11. Lopes, U.K., Valiati, J.F.: Pre-trained convolutional neural networks as feature extractors for tuberculosis detection. Comput. Biol. Med. **89**, 135–143 (2017)

12. Jaeger, S., et al.: Automatic tuberculosis screening using chest radiographs. IEEE Trans. Med. Imaging **33**(2), 233–245 (2013)

13. Hwang, S., Kim, H.E., Jeong, J., Kim, H.J.: A novel approach for tuberculosis screening based on deep convolutional neural networks. In: International Society for Optics and Photonics: Computer-Aided Diagnosis, vol. 9785, p. 97852W (2016)

14. Gabriella, I.: Early detection of tuberculosis using chest X-Ray (CXR) with computer-aided diagnosis. In: 2018 2nd International Conference on Biomedical Engineering (IBIOMED), pp. 76–79 (2018)

15. Antony, B., Nizar Banu, P.K.: Lung tuberculosis detection using x-ray images. Int. J. Appl. Eng. Res. **12**(24), 15196–15201 (2017)

16. Hussain, Z., Gimenez, F., Yi, D., Rubin, D.: Differential data augmentation techniques for medical imaging classification tasks. In: AMIA Annual Symposium Proceedings. AMIA Symposium, vol. 2017, pp. 979–984 (2018)

17. Tajbakhsh, N., et al.: Convolutional neural networks for medical image analysis: full training or fine tuning? IEEE Trans. Med. Imaging **35**(5), 1299–1312 (2016)

18. Saha, S.: A comprehensive guide to convolutional neural networks-the ELI5 way (2018)

19. Liu, C., et al.: TX-CNN: detecting tuberculosis in chest X-ray images using convolutional neural network. In: 2017 IEEE International Conference on Image Processing (ICIP), pp. 2314–2318 (2017)

20. Basheer, S., Jayakrishna, V., Kamal, A.G.: Computer assisted X-ray analysis system for detection of onset of tuberculosis. Int. J. Sci. Eng. Res. **4**(9), 2229–5518 (2013)

21. Le, K.: Automated detection of early lung cancer and tuberculosis based on X-ray image analysis. In Proceedings of the WSEAS International Conference on Signal, Speech and Image Processing, pp. 1–6 (2006)

22. Khutlang, R., et al.: Classification of Mycobacterium tuberculosis in images of ZN-stained sputum smears. IEEE Trans. Inf. Technol. Biomed. **14**(4), 949–957 (2009)

23. Alcantara, M.F., et al.: Improving tuberculosis diagnostics using deep learning and mobile health technologies among resource-poor communities in Peru. Smart Health **1**, 66–76 (2017)

24. Rohilla, A., Hooda, R., Mittal, A.: TB detection in chest radiograph using deep learning architecture. In: Proceeding of 5th International Conference on Emerging Trends in Engineering, Technology, Science and Management (ICETETSM-17), pp. 136–147 (2017)

25. Hamadi, A., Cheikh, N.B., Zouatine, Y., Menad, S.M.B., Djebbara, M.R.: Image-CLEF 2019: deep learning for tuberculosis CT image analysis. In: CLEF2019 Working Notes, vol. 2380, pp. 9–12 (2019)

26. Rajpurkar, P., et al.: Deep learning for chest radiograph diagnosis: a retrospective comparison of the CheXNeXt algorithm to practicing radiologists. PLoS Med. **15**(11), e1002686 (2019)

27. Nash, M., et al.: Deep learning, computer-aided radiography reading for tuberculosis: a diagnostic accuracy study from a tertiary hospital in India. Sci. Rep. **10**(1), 1–10 (2020)

# Dynamic Local Ternary Patterns for Gender Identification Using Facial Components

Salma M. Osman[1] and Serestina Viriri[2($\boxtimes$)] [iD]

[1] College of Computer Science and Information Technology,
Sudan University of Science and Technology, Khartoum, Sudan
[2] School of Mathematics, Statistics and Computer Science,
University of KwaZulu-Natal, Durban, South Africa
viriris@ukzn.ac.za

**Abstract.** Automated human face detection is a topic of significant interest in the field of image processing and computer vision. In the last years much emphasis has been on using facial images to extract gender of human which has become much used in many modern programs used in mobile phones. Local Binary Pattern (LBP), Local Directional Pattern (LDP) and Local Ternary Pattern (LTP) are popular appearance-based methods for extracting the feature from images. This paper proposes an improved Local Ternary Pattern technique called Dynamic Local Ternary Patterns to address the problem of gender classification using frontal facial images. The proposed Dynamic Local Ternary Patterns solve the limitation found in LBP (poorly performed in illumination variation and random noise) and LDP (produces inconsistence pattern), by increasing the number of the face components, applying the four cardinal directions namely North, East, West, South and using a dynamic threshold in LTP instead of a static one.

**Keywords:** Gender classification · Local ternary pattern · Local directional pattern · Local binary pattern · Dynamic local ternary pattern

## 1 Introduction

Over the last years, automated human gender detection has been a topic of significant interest in the field of computer vision. Texture analysis is the operation of machine learning to characterize texture in an image [1]. Texture can be described as a complex visual form composed of units or sub-patterns that has a wide variety of characteristics such as luminosity, color, size, slope, and shape. Texture is determined as the spatial variation of pixel intensities. Texture analysis is used in a variety of applications, including medical imaging, remote sensing and security.

© Springer Nature Switzerland AG 2020
L. J. Chmielewski et al. (Eds.): ICCVG 2020, LNCS 12334, pp. 133–141, 2020.
https://doi.org/10.1007/978-3-030-59006-2_12

Several feature methods have been proposed in literature, including Local Binary Pattern (LBP) [2], Local Directional Pattern (LDP) [3], Local Directional Pattern variance (LDPv) [4], Directional Local Binary Pattern (DLBP) [5] and many others. Face detection is a crucial first stage towards facial recognition technology and gender classification, which is used for identifying the face from the background and extracting the data [6]. There have been many face detection methods designed to identify faces within an arbitrary scene [7–12], the majority of these systems are only capable of detecting frontal or near-frontal views of the face.

The main idea behind using components is to compensate for pose changes by allowing a flexible geometrical relation between the components in the classification stage, however the main challenge is the selection of components; in addition, the face is more likely to be occluded and affected by abnormal lighting conditions. There has been a significant amount of research to date in various aspects of facial recognition, holistic-based methods, feature based methods and hybrid-based methods. Holistic-based methods perform well on images with frontal view faces and they are characterized by using the whole face image for recognition. However, they are computationally expensive as they require third-party algorithms for dimensionality reduction such as Eigen face techniques, which represents holistic matching of faces by the applications of PCA [13]. Feature-based approaches are much faster and robust against face recognition challenges. They are pure geometric methods; and they extract local features from facial landmarks [14]. Hybrid methods combines holistic and feature based methods to overcome the shortcomings of the two methods and give more robust performance.

In [15], a component based face recognition system is presented. It employs two-level Support Vector Machines (SVM) to detect and validate facial components. Learned face images are automatically extracted from 3-D head models that provide the expected positions of the components. These expected positions were employed to match the detected components to the geometrical configuration of the face and provide the expected positions of the components. These expected positions were employed to match the detected components to the geometrical configuration of the face.

Component-based face recognition studies are found at lower frequency in the literature. Even methods which compute similarity measures at specific facial landmarks, such as Elastic-Bunch Graph Matching (EBGM) [16] do not operate in a per-component manner. This work focuses on gender classification based on facial components analysis.

## 2    Methods and Techniques

The proposed methods and techniques for gender identification using facial components are described in the subsections below:

## 2.1 Dataset and Preprocessing

The LFW dataset consists of faces extracted from previously existing images and hence can be used to study recognition from images that were not taken for the special purpose of face recognition by machine. The LFW dataset images, in contrast contain arbitrary expressions and variation in clothing, pose, background, and other variables (Fig. 1).

**Fig. 1.** Sample of LFW dataset

The first stage of face recognition is face detection. It finds the facial area in image or video frame and passes it to the next stage. The region of interest is detected using Viola and Jones [21] face detection technique that searches the face portion.

Image pixel values converted to grey scale and the contrast is enhanced using histogram equalization with adaptive parameters. This is defined as follows: if we let $f$ be a given image represented as a $m_r$ by $m_c$ matrix of integer pixel intensities ranging from 0 to $L-1$, where L is the number of possible intensity values, often 256. Let p denote the normalized histogram of $f$ with a bin for each possible intensity.

## 2.2 Facial Components Detection

In component based gender identification, the most challenging task is to locate the components from the face. The Viola and Jones [22] algorithm is one of the powerful algorithms to perform this, although it does not cover all the components. The Discriminative Response Map Fitting (DRMF) model [24] is a fully automatic system that detects 66 landmark points on the face and estimates the rough 3D head pose. Figure 2 shows the eight component detected individually.

## 2.3 Feature Extraction

An important process in texture analysis is feature extraction. In this section, LBP, LDP together with the proposed DLTP are presented.

**Fig. 2.** Eight facial components detected

**Local Binary Pattern.** LBP [17] is a non-parametric descriptor that efficiently summarizes the structure of the image. Local Binary Pattern is considered to be tolerant of monotonic illumination changes and simple to compute. It is a texture analysis method but can also define local structures. Local Binary Pattern is computed by obtaining a decimal number for each pixel. Each pixel is then compared with the eight neighbouring pixels to the central pixel. The comparison is done by subtracting the central pixel from the neighbouring pixel, shown in Eq. (1). If the obtained value is negative then the pixel is assigned a 0; if a positive value is obtained, then that pixel is given a 1. The binary value is then read in a clockwise direction starting in the top left hand corner. The binary value that is obtained from each of these pixels is then converted to a decimal value. This decimal value that is obtained is then the pixel value, this value is between 0 and 255 pixel intensity. This can be calculated with the below equation:

$$LBP_{P,R}(x_c, y_c) = \sum_{p=0}^{p-1} s(i_p - i_c)2^p \tag{1}$$

$$S(X) = \begin{cases} 1, & if \quad x \geq 0 \\ 0, & if \quad x < 0 \end{cases}$$

where $i_c$ and $i_p$ are gray-level values of a central pixel $(x_c, y_c)$, P surrounding pixels are the circle neighbors with radius of R and s(x) is the defining 0 and 1 values. Nevertheless, The LBP is poorly performed in illumination variation and random noise in addition to high error rate when the background changes [18].

**Local Directional Pattern.** LDP [19] code is calculated by first applying Kirsch Masks $(m_{0,...,m_7})$ and eight-dimensional vector $(m_{0,...,m_7})$ can be calculated for the eight directions as follows:

$$C\left[\int (X,Y)\right] := (c_i = 1) \ if \ 0 \leq i \leq 7 \ and \ m_i \geq \psi \qquad (2)$$

where $\psi = k^{th}(M); M=\{m_0, m_1, ...m_7\}$

LDP code is then generated using the (k) most significant responses. Hence, the bits corresponding to the top (k) Kirsch Masks application responses are set to 1, and the remaining $(8-k)$ bits are set to 0. For a pixel at the position (X,Y), the LDP code $LDP_{x,y}$ is derived as be defined as:

$$LDP_{x,y}(m_0,...,m_7) = \sum_{i=0}^{7} S(m_i - m_k) \times 2^i \qquad (3)$$

where $m_k$ is the most prominent response. The histogram H is employed on the transformed image (LDP) to encode the image as a feature vector. The histogram obtained from the transformation can

$$H_i = \sum_{X=0}^{M-1} \sum_{Y=0}^{N-1} p(LDP_{x,y}, c_i) \qquad (4)$$

where $C_i$ is the $i_{th}$ ranked LDP value, for $i = 1, ..., \binom{8}{k}$ and p is defined as

$$p(x,a) = \begin{cases} 1 \ if \ x = a \\ 0 \ otherwise \end{cases} \qquad (5)$$

**Improved Local Ternary Pattern.** Dynamic LTP [20] solve the limitation found in LBP (poorly performed in illumination variation and random noise) and LDP (produces inconsistence pattern) by increasing the number of the face components, applying the four cardinal directions namely North, East, West, South and using dynamic threshold in LTP instead of static one.

| 0.0 | 0.72 | 0.89 |
|------|------|------|
| 0.21 | 0 | 0.44 |
| 1 | 0.93 | 0.03 |

| -1 | 1 | 1 |
|------|------|------|
| -1 | 0 | 0 |
| -1 | 1 | -1 |

(a) Normalized pixel    (b) Assigning DLTP code at
£= 0.5±0.1667

**Fig. 3.** Converted DLTP

**Parameter Selection.** Threshold $\epsilon$ is a set to 0.1667 deviation from 0.50 is selected as offset reference value for $\epsilon$ because it shows equal chance of there being an edge or not. The value of $\epsilon$ is chosen to ensure the probability is divided into 3 equal segments, one for each ternary bit. If the normalized response value is greater or equal to $0.5 + \epsilon$, its corresponding bit is set $+1$, if the normalized response value is less or equal to $0.5 - \epsilon$, its corresponding bit is set to $-1$, and the corresponding bit is set to the 0 if the normalized response is between $[0.5 - \epsilon$ and $0.5 + \epsilon$ as shown in Eq. (6) (Fig. 3).

The normalized responses are in the range of 0.0 and 1.0 which signifies the probability of edge from the central reference pixel stretching toward respective direction.

$$p(x_i) = \begin{cases} 1 & if \ x_i^{norm} \geq 0.05 + \epsilon \\ 0 & if \ 0.50 - \epsilon < x_i^{norm} < 0.50 + \epsilon \\ -1 & if \ x_i^{norm} \leq 0.05 - \epsilon \end{cases} \tag{6}$$

## 2.4   Classification

Classification involves taking the feature vectors extracted from the image and using them to automatically classify an images gender [23]. This is done by using different machine learning algorithms. In this research work, a SVM supervised machine learning algorithm was used to obtain the images gender. The supervised machine learning algorithm involves training the feature vector in order to compare an unknown feature vector with the trained data. The supervised machine learning algorithm used SVM.

**Support Vector Machine (SVM).** The SVM [24] is a trained algorithm for learning classification and regression rules from data. SVM is based on structural risk minimization and is related to regularization theory. The parameters are found by solving a quadratic programming problem with linear equality and inequality constraints. Using the kernel function allows for flexibility and can search for a wide variety in the dataset. The defined algorithm searches for an optimal separating surface, known as the hyper plane. All the data is then separated using the hyper plane. If there are too many outliers using the calculated hyper plane, a new hyper plane is calculated until the simplest hyper plane is formulated.

**The Evaluation of the System.** Cross-Validation is a statistical process of evaluating and comparing learning algorithms by dividing data into two parts: one used to learn or train a model and the other used to validate the model. Here we used 10 fold cross-validation which the data is randomly split into 10 equally (or nearly equally) sized segments or folds. Then 10 iterations of training and validation are performed such that within each iteration a different fold of the data is held-out for validation while the remaining 9 folds are used for learning [25].

# 3    Experimental Results and Discussion

In this study the data was collected from LFW frontal face databases which contain 500 image 250 for male and 250 female, also contain color and gray scale images. The performance of detect the eight regions of the face (eyes, nose, mouth, cheeks, chin and forehead) using Viola and Jones face detection technique and (DRMF) model, LBP, LDP and DLTP feature extraction techniques were applied and classifiers support vector machine (SVM) was used. Table 1 shows the gender identification accuracy from the individually extracted facial components, using four different feature extraction techniques. Table 2 shows the gender identification accuracy for combined facial components, using the proposed Dynamic Local Ternary Patterns.

**Table 1.** Gender identification accuracy per facial component using LBP, LDP, LTP and DLTP using SVM classifier

| | Feature extraction techniques | | | |
|---|---|---|---|---|
| Facial components | LBP | LDP | LTP | DLTP |
| Nose | 77.09% | 77.05% | 78.04% | 80.03% |
| Mouth | 86.09% | 85.35% | 87.50% | 87.90% |
| Eyes | 86.12% | 88.55% | 89.40% | 89.70% |
| Forehead | 87.32% | 90.35% | 90.90% | 91.10% |
| Chin | 78.12% | 81.55% | 81.89% | 84.02% |
| Cheeks | 82.12% | 82.50% | 83.95% | 84.40% |

**Table 2.** Gender identification accuracy using DLTP using SVM classifier

| Facial components | Accuracy (%) |
|---|---|
| Forehead + Eyes | 92.40% |
| Eyes + Mouth | 90.67% |
| Eyes + Nose + Mouth | 91.90% |
| Forehead + Eyes + Mouth | 98.90% |

Table 3 presents the comparison between the results achieved by related research studies on gender identification and our average gender identification rate. From the results, we can see that this study achieved better results than those of related works.

**Table 3.** Comparison of related gender identification results

| Method | Accuracy |
|---|---|
| LBP [26] | 90.10% |
| LDP [26] | 93.60% |
| LTP [26] | 93.70% |
| **Proposed DLTP** | **98.90%** |

## 4   Conclusion

This paper proposed an improved technique, Dynamic Local Ternary Pattern (DLTP) for accurate gender identification. The results achieved show that the whole face is not necessarily required for gender identification from facial images. The Discriminative Response Map Fitting model is used to detect the facial components (forehead, eyes, nose, cheeks, mouth and chin). The experimental results conducted on the LFW frontal face dataset, showed an accuracy rate of about 98.90%, and the most distinctive facial components are the forehead, eyes, and mouth. The proposed DLTP outperformed other related state-of-the-art feature extraction techniques, in terms of gender identification accuracy. For further work, it is envisioned that this model be extended to facial recognition in general and other facial characterization such age detection to determine the most distinctive and time-invariant facial components. Moreover, this research work can be extended to deep learning.

## References

1. Burrascano, P., Fiori, S., Mongiardo, M.: A review of artificial neural networks applications in microwave computer aided design (invited article). Int. J. RF Microw. Comput. Aided Eng. **9**(3), 158–174 (1999)
2. Wang, L., He, D.C.: Texture classification using texture spectrum. Pattern Recognit. **23**(8), 905–910 (1990)
3. Jabid, T., Kabir, M.H., Chae, O.: Local directional pattern (LDP) for face recognition. In: Proceedings of International Conference on IEEE on Consumer Electronics, pp. 329–330, January 2010
4. Hasanul Kabir, T.J., Chae, O.: Local directional pattern variance (LDPv): a robust feature descriptor for facial expression recognition. Int. Arab. J. Inf. Technol. **9**(4), 382–391 (2010)
5. Shabat, A.M., Tapamo, J.-R.: Directional local binary pattern for texture analysis. In: Campilho, A., Karray, F. (eds.) ICIAR 2016. LNCS, vol. 9730, pp. 226–233. Springer, Cham (2016). https://doi.org/10.1007/978-3-319-41501-7_26
6. Li, Y., Cha, S.: Face recognition system - arXiv preprint arXiv: 1901.02452 (2019). arxiv.org
7. Gunes, H., Piccardi, M.: A bimodal face and body gesture database for automatic analysis of human nonverbal affective behavior. In: International Conference on Pattern Recognition (ICPR), pp. 1148–1153 (2006)

8. Kobayashi, H., Tange, K., Hara, F.: Real-time recognition of six basic facial expressions. In: Proceedings of the IEEE Workshop on Robot and Human Communication, pp. 179–186 (1995)

9. Pentland, A., Moghaddam, B., Starner, T.: View-based and modular eigenspaces for face recognition. In: Proceedings of the IEEE Conference Computer Vision and Pattern Recognition, pp. 84–91(1994)

10. Rowley, H., Baluja, S., Kanade, T.: Neural network-based face detection. IEEE Trans. Pattern Anal. Mach. Intell. **20**(1), 23–38 (1998)

11. Sim, T., Baker, S., Bsat, M.: The CMU pose, illumination, and expression database. IEEE Trans. Pattern Anal. Mach. Intell. **25**(12), 1615–1618 (2003)

12. Viola, P., Jones, M.: Robust real-time object detection. In: International Workshop on Statistical and Computational Theories of Vision-Modeling, Learning, Computing, and Sampling (2001)

13. Turk, M.A., Pentland, A.P.: Face recognition using eigenfaces. In: IEEE Computer Society Conference on Computer Vision and Pattern Recognition 1991. Proceedings CVPR 1991, pp. 586–591, June 1991

14. Trigueros, D.S., Meng, L.: Face Recognition: From Traditional to Deep Learning Methods. arXiv:1811.00116v1 [cs.CV] 31 October 2018

15. Taheri, S., Patel, V.M., Chellappa, R.: Component-based recognition of faces and facial expressions. IEEE Trans. Affect. Comput. **4**(4), 360–371 (2013)

16. Bolme, D.S.: Elastic bunch graph matching (Doctoral dissertation, Colorado State University) (2003)

17. Domingos, P.: A few useful things to know about machine learning. Commun. ACM **55**(10), 78–87 (2012)

18. Heisele, B., Serre, T., Pontil, M., Poggio, T.: Component-based face detection. In: Proceedings of the IEEE Conference on Computer Vision and Pattern Recognition (CVPR) (2001)

19. Jabid, T., Kabir, M.H., Chae, O.: Robust facial expression recognition based on local directional pattern. ETRI J. **32**(5), 784–794 (2010)

20. Bashyal, S., Venayagamoorthy, G.K.: Recognition of facial expressions using Gabor wavelets and learning vector quantization. Eng. Appl. Artif. Intell. **21**(7), 1056–1064 (2008)

21. Du, H., Salah, S.H., Ahmed, H.O.: A color and texture based multi-level fusion scheme for ethnicity identification. In: SPIE Sensing Technology+ Applications, p. 91200B (2014)

22. Salah, S.H., Du, H., Al-Jawad, N.: Fusing local binary patterns with wavelet features for ethnicity identification. In: Proceedings of the IEEE International Conference on Signal Image Process, vol. 21, pp. 416–422 (2013)

23. Mitchell, T.M.: Machine Learning. WCB. McGraw-Hill, Boston (1997)

24. Hsu, C.W., Lin, C.J.: A comparison of methods for multiclass support vector machines. IEEE Trans. Neural Netw. **13**(2), 415–425 (2002)

25. Raniwala, A., Chiueh, T.: Architecture and algorithms for an based mulch-channel wireless mesh network. In: IIEEE Conference on Computer Communications, vol. 802, p. 11 (2005)

26. Ahmed, F., Kabir, M.H. (eds.): Consumer Electronics (ICCE), 2012 IEEE International Conference on Directional ternary pattern (DTP) for facial expression recognition. IEEE, Las Vegas (2012)

# Depth Perception Tendencies in the 3-D Environment of Virtual Reality

Jana Polláková, Miroslav Laco[(✉)][iD], and Wanda Benesova

Slovak University of Technology in Bratislava,
Ilkovičova 2, 842 16 Bratislava, Slovakia
{miroslav.laco,vanda_benesova}@stuba.sk
https://vgg.fiit.stuba.sk/

**Abstract.** The human brain is not able to process the vast amount of visual information that originates in the environment around us. Therefore, a complex process of human visual attention based on the principles of selectivity and prioritization helps us to choose only the most important parts of the scene for further analysis. These principles are driven by the visual saliency derived from certain aspects of the scene parts. In this paper, we focus on the objects' depth salience which undoubtedly plays its role in processing visual information and has still not been thoroughly studied until now.

The aim of our work is to investigate depth perception tendencies using an advanced experimental methodology in the environment of virtual reality. Based on the state-of-the-art of the attention modelling in the 3-D environment, we designed and carried out an extensive eye-tracking experimental study in the virtual reality with 37 participants observing artificial scenes designed for exploring trends in the depth perception in the virtual 3-D environment. We analyzed the acquired data and discuss the revealed depth perception tendencies in virtual reality alongside future possible applications.

**Keywords:** Visual attention · Depth perception · Virtual environment · Saliency

## 1 Introduction

Human visual attention constantly selects the most important visual information from the environment as the amount of visual information incoming from around us is enormous. This optimization process is based on selectivity, which directly depends on the saliency of visual stimuli at the scene. There are many aspects affecting the saliency, e.g. colour, shape, orientation, motion, and many more [2]. Each of the aspects affects our perception in a very unique way. However, some of the perception tendencies have certain common phenomena (e.g. *pop-out effect* [7]) which can be further modelled to simulate visual attention as an optimization process in many application fields.

© Springer Nature Switzerland AG 2020
L. J. Chmielewski et al. (Eds.): ICCVG 2020, LNCS 12334, pp. 142–150, 2020.
https://doi.org/10.1007/978-3-030-59006-2_13

Some of the aspects affecting human perception have not been thoroughly discovered, or examined, yet. This offers wide space for further research of computer scientists in the field of attention research and modelling. One of the most recent urges in the field of attention modelling is to investigate the attention behaviour in the 3-D environment. Due to the technical limitations, this was not possible to be done ever before. Thanks to the technical progress and possibilities, we are able to examine the perception tendencies under various conditions and in a wide variety of environmental conditions - e.g. in the real world [8,9,11]. In this work, we focused on exploring the depth perception tendencies in the 3-D environment of virtual reality. We come out of the work of Laco et al. and assume that human visual attention in the virtual reality would follow the same perception tendencies as in the real environment.

Previous studies have brought a broad range of knowledge into stereoscopic visual depth-bias in the artificial 3-D environment, as well as monocular depth-bias while perceiving a 2-D visual scene [1,9]. Our motivation is to contribute to the knowledge of the research community with novel findings related to the depth perception tendencies in the environment of virtual reality to open new attention modelling possibilities in the future.

## 2 Related Work

Visual attention modelling has been widely studied during the past decades. There were introduced dozens of models which are built on top of various features and aspects of the attention and use a wide variety of image processing and computer vision principles to fulfill the goal to predict human eye behaviour [2]. Throughout the past years, machine learning principles in visual attention modelling and attention modelling using deep neural networks became very popular. Such models instantly grabbed the front-most position in famous saliency benchmarks [3] (e.g. the MIT saliency benchmark [4]). However, the attention models are still built on top of the datasets from the eye-tracking experimental studies where certain aspects of human attention behaviour are explored. The technical possibilities allowed us to study the attention in the unconstrained 3-D environment only a few years ago. Therefore, there are only a few studies related to the visual perception tendencies in virtual reality, yet.

### 2.1 Visual Attention Studies in the Virtual Reality

Marmitt et al. [10] created a novel metric for comparing human and artificial scan-paths in virtual reality. Moreover, they analyzed the behaviour of common attention models in a virtual environment. Their results revealed the significantly worse performance of these common models when applied in the environment of virtual reality.

Sitzmann et al. [14] focused their research on visual attention in virtual environments and specifically on how users explore the virtual scenes. Oppositely to Marmitt et al., they revealed a great similarity between scan-paths of observers in the virtual environment and the scan-paths from observing a common 2-D visual scene from the display. There is no other research work exploring the depth perception tendencies in the virtual reality, to our knowledge, yet.

## 2.2   Depth Perception in the 3-D Environment

There are certain aspects that strongly affect the attention in the 3-D environment. One of them is the depth of the scene which became a subject of the latest research, as well. Roberts et al. [13] and Desingh et al. [6] concurred that depth bias has a significant impact on the detection of salient parts at the scene and can contribute to computational saliency models to improve the prediction. The study of Wang et al. [1] proved the significant influence of depth on human visual attention while viewing stereoscopic visual stimuli. The evaluated experimental study showed that participants tended to focus their attention on the nearest objects of visual scenes sooner than on the other objects. However, the experimental study of Polatsek et al. indicated the opposite - the farthest objects turned out to be the most salient ones during their experimental study in the natural 3-D environment [12]. A similar research approach was proposed by Laco et al. [9] and his eye-tracking study with projecting captures of the scenes from Polatsek's experiment on a wide-screen display. The results of this experimental study did not support the pop-out effect hypothesis, as in this type of scenes the fixation density was almost equally distributed among all objects. Furthermore, neither the assumption that closer objects would greatly attract human attention was proven.

We decided to further verify the hypothesis of Laco et al. [9] and Polatsek et al. [12] in the virtual environment to extend their findings and provide another insight into depth perception in the three-dimensional environment. There is no eye-tracking dataset that could be used for such purposes, to our knowledge, yet. Therefore, we proposed our own experimental study in the environment of virtual reality according to the proposals of the Laco et al. [9].

## 3   Experimental Study Proposal

Our proposed experimental study followed the same methodology and scene configurations as in Laco et al. [9] to maintain the compatibility of the results for the future comparison. The prepared visual scene provided to the participants of the experiment consisted of 8 identical objects - white balls - situated right in front of a viewer. White balls were chosen as the most visually neutral stimuli to avoid interference of the depth perception tendencies with other aspects affecting visual salience. The objects were arranged into the shape of the octagon to avoid the strong effect of the visual center-bias [3] and labelled **A-H** to further reference them (see Fig. 1).

According to the proposal of Laco et al., the objects could appear at the scene in 6 different depth levels - from 245 cm up to 395 cm from the viewer (see Fig. 1).

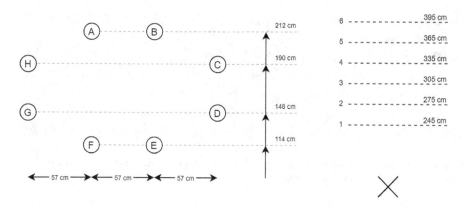

**Fig. 1.** Left: Frontal schematic view of the object configuration in the virtual space alongside with the object labels. Right: Schematic bird-view of the virtual scene configuration displaying the depth-levels where the objects could be placed

**Table 1.** Table of the object depth-level configurations. Each configuration is represented by sorted depth levels belonging to the specific object in the order **A-H** [9, 12]

| Scene type | Depth levels | Depth level steps between adjacent objects | Variant | | |
|---|---|---|---|---|---|
| | | | a | b | c |
| I | 2 | The only one step of 1 | 12111111 | 11121111 | 11111112 |
| II | 3 | Alternating zero steps and steps of 2 | 33553311 | 53311335 | 31133443 |
| III | 4–5 | Only steps of 1 | 34565654 | 32123454 | 56543234 |
| IV | 4–6 | Steps of 1, but one step of 3 | 45632123 | 52321234 | 54565434 |
| V | 5–6 | Random - steps vary from 0 up to 4 | 62543134 | 32456321 | 34662124 |

**Fig. 2.** The capture of the configuration of the objects in scene type **I** variant **a**

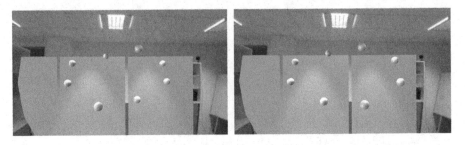

**Fig. 3.** The capture of the configuration of the objects in scene type **IV** variant **b** (left) and scene type **V** variant **a** (right)

The visual scenes for the eye-tracking experimental study were divided into 5 scene types which were intended to include the most common depth configurations of the objects at the natural scenes (e.g. by maintaining various steps among neighbour objects, various depth levels present at one scene, etc.). There were 3 different variations of each scene type to avoid biasing the eye-tracking measurements by bounding them to the specific scene configuration. The generated scene configurations can be found in Table 1. Capture of the experimental virtual scene with the objects in various configurations is displayed in Fig. 2 and 3.

### 3.1 Participants and Apparatus

The sample of participants consisted of 37 students aged from 21 to 23. The sample of the participants was without any visual impairment or disorder.

The apparatus used for the eye-tracking studies was the headset for the virtual reality *HTC Vive* with an additionally installed binocular 200 Hz add-on eye-tracker from *Pupil Labs*. The virtual scene was designed and created using the Unity development platform for video games, 2D, 3D, augmented and virtual reality visualizations.

### 3.2 Methodology of the Experimental Study

The whole experimental study took place in the laboratory where only the participant and the instructor were present to avoid the presence of any disturbing elements biasing the attention. Each participant was asked to mount a headset on their head. Afterwards, the eye-tracker inside the headset was calibrated to adjust the eye-tracking precision using seven-point calibration designed by the *Pupil Labs*. After proper calibration, the participant was exposed to the initial scene setup with all the objects in the same depth. The calibration was verified on this scene and participants got familiar with the scene itself, so their attention was further affected only by changes in the object's depth. Participants were instructed to look freely (free-viewing task) and then exposed to randomly shuffled 15 different scene configurations for 10 s each. The eye-tracking data were measured during the whole session.

## 3.3   Methodology for Evaluation

We collected the eye-tracking data from 37 recording sessions with all the participants and stored the data in a standardized format referencing the recording timestamp, specific type and variant describing the observed visual scene and the label of the fixated object if any (label **A-H**)[1].

For the evaluation purposes, we used a fixation density metric (see Eq. 1). This metric was chosen for the data analysis as the most suitable as it is often used for evaluating eye-tracking data from similar experimental studies (see Bylinskii et al. [5]). The fixation density metric is based on summing up the relative fixation duration on an object in certain depth as follows:

$$d_i = \frac{\sum t_i}{p}, \tag{1}$$

where $d_i$ is fixation density on a certain object of interest with the label $i$, and $t_i$ is the duration of the fixation on the object of interest with the label $i$ within the given period of time given for the scene observation $p$ (10 s) [9].

## 4   Evaluation and Discussion

We evaluated the obtained data from our eye-tracking experimental study and confronted the results with the two hypotheses about attention behaviour in the 3-D environment following the previous work described in Sect. 2. The first hypothesis posed in Laco et al. and originating from the work of Itti et al. states that visual *pop-out effect* should be present when perceiving depth in the 3-D environment. In other words, if the object stands-out in the depth channel of the 3-D scene, it should be more salient than the others. The second hypothesis follows the revealed depth perception tendencies about higher salience of the closest objects at the scene. We hypothesize that these trends should be also present in the environment of virtual reality. For the discussion purposes, we visualize the obtained results in the box-plots where the normalized fixation ratio is grouped by the depth levels in which the objects appeared at.

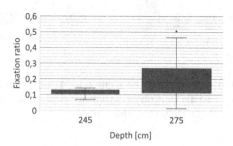

**Fig. 4.** Fixation distribution regarding the scene type **I**, variant **a** demonstrating the depth *pop-out* effect in the conditions of virtual reality. The single object in the depth of 275 cm grabbed generally the most of the observers' attention

---

[1] The dataset is publicly available on demand.

## 4.1  Depth Pop-Out Effect

The motivation behind the object configurations of the first scene type was to explore if one object appearing in different depth level than the others would grab the most of the observers' attention - the so-called *pop-out effect* in the depth channel. The *popping-out* object was always the one further away from the observer (depth of 275 cm). We examined evaluation results for all three variants of the scene type **I** and we discovered a noticeable difference in fixation densities between the depth level of the *popping-out* object and the depth level of the other objects (see Fig. 4). Therefore, we assume that the *pop-out* effect in the depth channel is present in the environment of virtual reality as well as if perceiving the 2-D visual scene. Thus, we support our first hypothesis.

## 4.2  The Most Salient Depth

All the other scene types described in Sect. 3 were arranged to cover most of the depth configurations of the objects in natural scenes. They varied greatly in the depth configurations (see Table 1 and Fig. 3). However, the perception tendencies bound to all the scene types and variants revealed very similar depth perception trends. We can clearly notice the pattern in the salience of the objects in certain depth levels in Fig. 5. Surprisingly, the revealed pattern integrates both types of contrary conclusions on the depth perception in the 3-D environment described in Sect. 2. More specifically, closer, as well as farther objects from the observer, tend to be more salient in the virtual reality than the objects in the middle of the depth range of the visual scene. Therefore, we consider our second hypothesis as partially supported by the results of our experimental study.

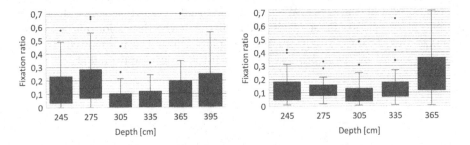

**Fig. 5.** Fixation distribution regarding the scene type **4**, variant **b** (left) and the scene type **V**, variant **a** (right) reveals that objects that are closer and farther away from the viewer tend to be more salient than the other ones

We assume that such revealed trends may be the reason for surprisingly worse-performing attention models, built on top of the 2-D eye-tracking patterns, applied in the 3-D environment of the virtual reality (e.g. by Marmitt et al. [10]). We claim that human visual attention and perception patterns should be thoroughly studied also in the environment of the virtual reality to prove some already explored attention patterns bound to observing 2-D stimuli and to bring more insight on how to build up the attention models for the 3-D virtual environment.

### 4.3 Limitations

Our experimental study was designed to be held in specific conditions in a copy of the virtual laboratory proposed by Laco et al. [9] where the depth range of the visual scene was 4 meters. This limitation was set for the future possible comparison of the results from both studies. Therefore, the evaluation results of this study cannot be taken as general trends in the depth perception and are bound to the indoor virtual environment with a small depth range.

## 5   Conclusion

We designed and carried out an extensive eye-tracking experimental study with 37 subjects based on the proposal of the state-of-the-art work in the field of visual attention research. The study was aimed to analyze the impact of the visual depth-bias on visual attention of the observers in the 3-D environment in virtual reality. The results of our eye-tracking experimental study revealed interesting trends in the visual depth-bias which could be used for a better understanding of the depth perception tendencies in virtual conditions. Our findings interconnect the majority of the previous studies and support both of the most popular assumptions regarding depth perception - namely high visual saliency of both the closest and farthest objects in the scene. We believe that our study will offer the opportunity for a further and thorough analysis of visual attention trends in the environment of the virtual reality in the future.

In future work, we would like to verify the revealed depth perception tendencies in real virtual reality applications. Moreover, we would like to improve the attention models with the revealed trends for more accurate attention predictions in the virtual reality.

## References

1. Wang, J., et al.: Study of depth bias of observers in free viewing of still stereoscopic synthetic stimuli. J. Eye Mov. Res. **5**(5), 11 (2012)
2. Borji, A., Itti, L.: State-of-the-art in visual attention modeling. IEEE Trans. Pattern Anal. Mach. Intell. **35**(1), 185–207 (2013)
3. Borji, A., Cheng, M., Hou, Q., et al.: Salient object detection: a survey. Comput. Vis. Media **5**, 117–150 (2019). https://doi.org/10.1007/s41095-019-0149-9
4. Bylinskii, Z., et al.: MIT saliency benchmark (2015)
5. Bylinskii, Z., Judd, T., Oliva, A., Torralba, A., Durand, F.: What do different evaluation metrics tell us about saliency models? IEEE Trans. Pattern Anal. Mach. Intell. **41**(3), 740–757 (2018)
6. Desingh, K., Krishna, K.M., Rajan, D., Jawahar, C.: Depth really matters: improving visual salient region detection with depth. In: BMVC (2013)

7. Itti, L., Koch, C.: Computational modelling of visual attention. Nat. Rev. Neurosci. **2**(3), 194 (2001)
8. Laco, M., Benesova, W.: Depth in the visual attention modelling from the egocentric perspective of view. In: Eleventh International Conference on Machine Vision (ICMV 2018). Proceedings of SPIE (2019). https://doi.org/10.1117/12.2523059
9. Laco, M., Polatsek, P., Benesova, W.: Depth perception tendencies on a widescreen display: an experimental study. In: Already Submitted to: Twelfth International Conference on Machine Vision (ICMV 2019). Proceedings of SPIE (2019)
10. Marmitt, G., Duchowski, A.T.: Modeling visual attention in VR: measuring the accuracy of predicted scanpaths. Ph.D. thesis, Clemson University (2002)
11. Olesova, V., Benesova, W., Polatsek, P.: Visual attention in egocentric field-of-view using RGB-D data. In: Ninth International Conference on Machine Vision (ICMV 2016), vol. 10341, p. 103410T. International Society for Optics and Photonics (2017). https://doi.org/10.1117/12.2268617
12. Polatsek, P.: Modelling of human visual attention. Dissertation thesis, FIIT STU, Bratislava (2019)
13. Roberts, K.L., Allen, H.A., Dent, K., Humphreys, G.W.: Visual search in depth: the neural correlates of segmenting a display into relevant and irrelevant three-dimensional regions. NeuroImage **122**, 298–305 (2015)
14. Sitzmann, V., et al.: Saliency in VR: how do people explore virtual environments? IEEE Trans. Visual Comput. Graphics **24**(4), 1633–1642 (2018)

# Optimisation of a Siamese Neural Network for Real-Time Energy Efficient Object Tracking

Dominika Przewlocka[ID], Mateusz Wasala[ID], Hubert Szolc[ID],
Krzysztof Blachut[ID], and Tomasz Kryjak[✉][ID]

Embedded Vision Systems Group, Computer Vision Laboratory,
Department of Automatic Control and Robotics,
AGH University of Science and Technology, Krakow, Poland
{dprze,wasala,szolc,kblachut,tomasz.kryjak}@agh.edu.pl

**Abstract.** In this paper the research on optimisation of visual object tracking using a Siamese neural network for embedded vision systems is presented. It was assumed that the solution shall operate in real-time, preferably for a high resolution video stream, with the lowest possible energy consumption. To meet these requirements, techniques such as the reduction of computational precision and pruning were considered. Brevitas, a tool dedicated for optimisation and quantisation of neural networks for FPGA implementation, was used. A number of training scenarios were tested with varying levels of optimisations – from integer uniform quantisation with 16 bits to ternary and binary networks. Next, the influence of these optimisations on the tracking performance was evaluated. It was possible to reduce the size of the convolutional filters up to 10 times in relation to the original network. The obtained results indicate that using quantisation can significantly reduce the memory and computational complexity of the proposed network while still enabling precise tracking, thus allow to use it in embedded vision systems. Moreover, quantisation of weights positively affects the network training by decreasing overfitting.

**Keywords:** Siamese neural networks · DNN · QNN · Object tracking · FPGA · Embedded vision systems

## 1 Introduction

Visual object tracking is one of the most complex tasks in computer vision applications and despite the great number of algorithms developed over the last decades, there is still plenty of room for improvement. Well known challenges, typical for this task, involve object appearance changes (size, rotations), variable dynamics and potential similarity to the background, illumination changes, as well as object disappearing. To handle the enlisted situations, advanced algorithms have to be used. On the other hand, such systems are often used in

© Springer Nature Switzerland AG 2020
L. J. Chmielewski et al. (Eds.): ICCVG 2020, LNCS 12334, pp. 151–163, 2020.
https://doi.org/10.1007/978-3-030-59006-2_14

edge devices like smart cameras used for video surveillance and vision systems for autonomous vehicles (cars, drones). This rises another challenges: real-time processing and energy efficiency requirements, as well as support for high resolution video stream – Full High Definition (FHD - 1920 × 1080) and Ultra High Definition (UHD – 3840 × 2160).

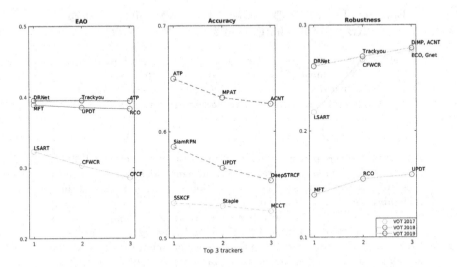

**Fig. 1.** EAO (Expected Average Overlap), Accuracy and Robustness of the best 3 trackers from three last editions of VOT: 2017, 2018 and 2019 in the baseline category

Following the subsequent editions of the well known in the community VOT (Visual Object Tracking) challenge, one can observe a noticeable progress in accuracy of the proposed trackers over the last few years. In the VOT challenge the trackers are evaluated using accuracy, robustness and the expected average overlap (EAO) which combines the two previous factors to reasonably compare different algorithms. Figure 1 shows results of the top 3 trackers from the last three editions of the challenge in the baseline category. From 2017 to 2019 the EAO value of the best tracker improved by 18.23%, which has to be considered as a significant progress.

Many recently proposed solutions either do not operate in real-time or are accelerated on rather energy inefficient GPUs. Their accuracies places them high in VOT (Visual Object Tracking) challenges, thus makes them appropriate for applications with a great need for precise tracking, such as autonomous vehicles (AV), advanced driver assistance systems (ADAS) or advanced video surveillance systems (AVSS). There is therefore a necessity for using them on edge or embedded devices with limited power budget.

However, the computational complexity of the state-of-the-art trackers is so demanding, that in order to achieve real-time processing, high-end GPUs are needed. For many applications this is often unacceptable, since GPUs consume a significant amount of energy: for example autonomous vehicles powered by accumulators need low-power solutions working in real-time. There is therefore a need for lightweight algorithms, in terms of memory and computational complexity, that are possible to accelerate on embedded devices. B It is obvious that such trackers still need to provide the best possible accuracy, so research in this area has to focus on optimisation of the state-of-the-art solutions. One of the possible choices for the target computing platform are FPGAs (Field Programmable Gate Arrays) or heterogeneous devices (combining programmable logic and CPUs), as they enable high level of parallelisation, while consuming relatively low energy (a few watts).

Lately, many of the best trackers use Siamese Neural Networks. A Siamese network is a Y-shaped network with two branches joined to produce a single output. The idea itself is not new, as it originated in 1993 in works about fingerprint recognition [15] and signature verification [16]. A Siamese network measures the similarity of two processed inputs, thus it can be considered as a similarity function. In fact, many of the Siamese-based trackers rely solely on that assumption. The exemplar image of an object (from the first frame) is treated as the first input to the network, while the region of interest (ROI), where the object might be present in the consecutive frame, forms the second one. The network searches for similar to the object's appearance areas in the search ROI. Depending on specific tracker, the output is a new bounding box or just the location of the object (the centre of mass) that has to be further processed to obtain the bounding box.

The presented research focuses on adapting a state-of-the-art tracker based on Siamese Neural Networks to embedded devices. We aim to optimise the network for memory and computational complexity which shall directly translate into meeting real-time requirements via an FPGA implementation. The main contribution of this paper is the research on optimising Siamese Neural Networks for object tracking, by reducing the computational and memory complexity. To the authors' best knowledge no prior work described the results of such various quantisations – from uniform integer quantisation of 16, 4 bits, ternary and binary filters.

The rest of the paper is organised as follows: Sect. 2 briefly describes previous work in related domains – deep learning, tracking and acceleration of computations. Section 3 describes tracking using Siamese Neural Networks. In Sects. 4 and 5 tools and optimisation techniques for reducing the computational and memory complexity of neural networks are presented. Ultimately, Sect. 6 describes the conducted experiments and Sect. 7 the obtained results. Finally, in Sect. 8 the presented research is summarised and possible future work is discussed.

## 2    Previous Work

Recently several trackers based on Siamese Neural Networks were proposed by the scientific community. In [1] the authors proposed a tracker based on a Siamese Neural Network named GOTURN. The two branches are based on the Alexnet

[21] DCNN and the network is trained for bounding box regression. Given two input images, one as an exemplar of object to track and the second with the ROI, the network predicts the location of the object in the new frame. This algorithm was implemented using Nvidia GeForce GTX Titan X GPU with cuDNN acceleration and achieved the speed of 6.05 ms per frame (165 fps) and GTX 680 GPU with the speed of 9.98 ms (100 fps). The tracker is able to process 2.7 frames per second on a CPU.

In [2] the authors proposed a slightly different solution. Based on the two inputs (exemplar and instance – ROI), the Siamese Network estimates the location of object in the ROI. The possible variance of the object's size is handled by analysing multiple scales, thus calculating the network's output multiple times for a single frame. Tracking implemented on computer with Nvidia GeForce GTX Titan X and an Intel Core i7-4790K 4.0 GHz runs at 86 58 fps with respectively 3 and 5 scales.

The direct continuation of the work [2] resulted in [5]. The authors presented a tracker which combines a Siamese Neural Network and correlation filters. The results showed that it allowed to achieve state-of-the-art performance with lightweight architectures. Using the same hardware, the tracker achieves a speed of 75–83 fps (depending on network's depth).

In [3] the authors presented a Dynamic Siamese Network that allows for online learning of the target appearance from previous frames. The network's output is a score map (similar as in [2]) that indicates the centre of the object's location. The tracker 45 fps on a Nvidia Titan X GPU.

Another interesting idea was presented in [6], where the network is trained for bounding box regression but without any prior knowledge abut the bounding box (there are no pre-defined anchor boxes nor multi-scale searching). This solution achieved a speed 40 fps on Nvidia GTX 1080Ti and Intel Xeon(R) 4108 1.8GHz CPU with AUC equals to 0.594 on OTB100 benchmark.

A comparison of the above-mentioned Siamese-based trackers evaluated on OTB 2013 benchmark is presented in Table 1. Moreover, based on the results from the VOT 2019 challenge [4] 21 (37%) of participating trackers were based on Siamese Networks, from which 3 made it to the top ten best trackers according to the EAO measure.

**Table 1.** Comparison of trackers based on Siamese Neural Networks. The choice of the benchmark OTB 2013 was dictated by reports in articles provided by the authors.

| Tracker | AUC in OTB 2013 | Speed (fps) | Acceleration |
|---|---|---|---|
| GOTURN [1] | 0.447 | 165 | Nvidia Titan X GPU |
| SiamFC-3s [2] | 0.607 | 86 | Nvidia Titan X GPU |
| DSiam [3] | 0.642 | 45 | Nvidia Titan X GPU |
| CFNet-conv2 [5] | 0.611 | 75 | Nvidia Titan X GPU |

All the above mentioned algorithms have high tracking accuracy and operate in real-time with the use of high-end GPUs. Hence, if one wishes to use them in low energy systems, implementation on embedded devices is necessary. One of the possible solutions is to use FPGA devices, as they provide the possibility of computation parallelisation and are energy efficient. The currently proposed solutions have high memory and computational complexity and usually it is impossible to fit such networks entirely on the considered device. Regardless of tracking, several simplifications for convolutional neural networks were proposed: binary or XNOR networks with zero-one weights and activations [7], ternary ($\{-1, 0, 1\}$) networks [8], networks with other quantisations (uniform integer quantisation) adopted a priori [9], as well as trained [10], or pruning, i.e. reducing networks' parameters by removing neurons or connections between them [11]. Moreover, recently there is a strong trend for research on efficient network design [12].

There are few papers covering acceleration of Siamese Neural Networks for tracking. In [13] the authors presented a tiny Siamese Network for visual object tracking, that due to the low computational complexity can be implemented in embedded devices. The presented algorithm operates with the speed 16 fps on a CPU 129 fps on a GPU and is fairly comparable to other state-of-the-art solutions based on Siamese Neural Networks (like SiamFC or GOTURN [1,2]) by achieving the score of AO (average overlap) equal to 0.287 (for example, SiamFC achieves 0.349 AO 39 fps on a GPU).

Finally in [14] the authors presented an FPGA implementation of lightweight system for visual object tracking using a Siamese Neural Network. Basing on the [2] solution, the authors developed a network with a more regular architecture (with filters of the same kernel sizes) and then performed pruning and quantisations. This allowed to achieve results comparable, in context of precision or AUC, to the original solution. The tracker operates with the speed 18.6 fps on a ZedBoard and is highly energy efficient with only 1.284W.

Despite the great interest in the area of tracking based on Siamese Neural Networks, not much research is concentrated on accelerating this excellent trackers in order to use them in systems with low power requirements. Therefore, we believe that the results of our experiments fill a gap in this domain.

## 3  Tracking with Siamese Neural Networks

### 3.1  Siamese Neural Networks

A Siamese network is a Y-shaped network with two branches joined to produce a single output. Irrespective of the branches' structure, a Siamese network can be considered as a similarity function that measures the resemblance between two inputs. We can define the basic Siamese architecture using Eq. (1), where $\phi$ stands for extracting deep features (e.g. via network's branches), $z$ represents the template (e.g. exemplar, object to track), $x$ represents the patch that is compared to the exemplar (e.g. search region, ROI) and $\gamma$ is a cross-correlation.

$$y = \gamma(\phi(z), \phi(x)) \tag{1}$$

## 3.2  Tracking

As discussed in Sect. 2, several trackers based on Siamese Networks exist. In this research we focus on the solution presented firstly in [2] and then modified in [5], mainly due to its good performance and simplicity (which is important in context of FPGA implementation). Its scheme is presented in Fig. 2. The input $z$ represents the tracked object selected in the first frame and scaled to $127 \times 127$ pixels. In the next frames, the ROI around the previous object's location is selected and presented to the network as the input $x$, scaled to $255 \times 255$ pixels. Both patches (exemplar and ROI) result in two blocks of feature maps, of sizes $17 \times 17 \times 32$ (for $z$) and $49 \times 49 \times 32$ (for $x$) respectively. Finally they are cross-correlated and a heat map of size $33 \times 33$ is obtained. Its highest peak indicates the object's location. The ROI is selected approximately 4 times greater than object's size, over five scales to handle object's size variations.

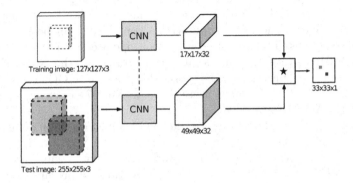

**Fig. 2.** A fully-convolutional Siamese Network for tracking [2]

## 4  The Used Tools: Brevitas and FINN

In order to implement deep learning algorithms in FPGA devices it is highly advisable to somehow optimise the network's architecture. This increases the performance and, in some cases, even allows to only use internal memory resources. Apart from designing lightweight networks, there are several other options to compress it's parameters, like quantisation and pruning. However, these methods are not independent of the training process. For a long time most popular tools for deep learning, such as Tensorflow or Pytorch standalone, did not contain such extensions and no open source tool was able to perform such optimisations. However, as the acceleration of deep networks on embedded devices gained on popularity in last few years, this situation changed. This resulted in several interesting tools for deploying them on FPGA/SoC platforms (Field Programmable Gate Array/System on Chip), e.g. Caffe to Zynq from Xilinx reVISION stack, Vitis AI or open source Brevitas/FINN. In this work we will focus on the last solution.

Brevitas and FINN are complementary tools, but as they are still under development and not all features are yet ready. Brevitas is a Pytorch library that enables the training of quantised networks, so it allows optimising the considered algorithm. The trained model can be then compiled into a hardware design using the FINN tool. However the Brevitas-to-FINN part of the flow is currently available for a very constrained set of network layers. The current version of the framework is sufficient for designing algorithms ready for hardware implementation in the future releases of FINN or "manually".

## 5    Optimisation Methods

### 5.1    Quantisation

Quantisation allows to reduce the number of bits needed to store values, which can be (and often is) critical for deploying networks on embedded devices. Both weights and activations can be quantised. Despite several approaches involving quantising weights after training, the best results are obtained when training the network with reduced precision parameters. Using Brevitas, one can set both type of quantisation (uniform integer quantisation, binary or ternary), as well as particular bit width for weights and activations.

### 5.2    Pruning

The idea of pruning is as old as 1990, when Yan LeCun [19] showed that removing unimportant weights from a neural network can result in better generalisation, fewer training examples needed and acceleration of training, as well as inference. A neural network is a rather redundant structure and by finding the neurons that do not contribute much to the output, its architecture can be optimised. One possible way to do that is to rank the neurons and search for those which are less relevant – according to L1/L2 norm of weights in a neuron, mean activations and other measures. In practice, the process of pruning is iterative with multiple fine-tuning steps: after suppressing low ranked neurons, the network's accuracy drops and a retraining step is necessary. Usually the network is pruned several times, deleting only a relatively small number of neurons at once.

## 6    Experiments

The proposed network is similar to the one presented in [2]. In order to optimise it, several quantisation experiments were performed. It was decided to focus more on the hidden layers than on the input and output ones, as while conducting preliminary tests it turned out that radical quantisations of all layers causes difficulties in training. Using weights with higher precision for the last layers results in better accuracy of tracking, which is showed in experiments. Moreover, based on similar observations, the activation layers also operate with higher precision (with uniform integer quantisation).

## 6.1  Training

In order to obtain representative and comparable to each other results, we have defined the same training parameters for all test cases. The networks were trained using the ILSVRC15 dataset. For the presented experiments, approximately 8% (that is 10000 image pairs of object and ROI) of the dataset was used for training. The authors of original Siamese Network compared the accuracy – measured as average IOU – of their tracker while training on different portions of the ImageNet dataset. In that case the 8% evaluation resulted in an 48.4% accuracy (for the whole dataset: 52.4%). Thus, to reduce the training time, we have followed the same approach – in our case one training experiment took about 12 h on a PC with Nvidia GeForce GTX 1070 GPU. The number of epochs was set to 30. Naturally, when training the network one should adjust its architecture to the task and available data to prevent the network from learning "by heart". Nevertheless, in this case, when deliberately training on a small subset of the dataset, this could happen, thus number of epochs was minimised to the value that allowed for a decent evaluation while preventing from overfitting. After these preliminary experiments, the whole process can be repeated for a one chosen network (with the best relation of computational and memory complexity to accuracy) on the whole dataset to obtain better results.

**Table 2.** Quantisations of the proposed Siamese Neural Network for tracking. FP stands for floating point precision, INT for uniform integer quantisation, TERNARY and BINARY respectively for two and one bits precision

| No | First layer | Hidden layers | Last layer | Activation | Convolutional parameters |
|----|-------------|---------------|------------|------------|--------------------------|
| 1  | FP 32       | FP 32         | FP 32      | FP 32      | 85.5 MB                  |
| 2  | INT 16      | INT 16        | INT 16     | INT 32     | 38.3 MB                  |
| 3  | INT 16      | INT 4         | INT 8      | INT 32     | 13.4 MB                  |
| 4  | INT 16      | TERNARY       | INT 8      | INT 32     | 9.3 MB                   |
| 5  | INT 16      | BINARY        | INT 8      | INT 32     | 7.2 MB                   |

The conducted experiments are summarised in Table 2. They were selected to evaluate how the reduced precision affects the training progress and it's results. The first case represents the network with floating point precision and is treated as the baseline. The last column shows the memory needed for convolutional filters (other parameters are constant for all test cases). It is worth noting that on FPGA devices the use floating point computations is generally more complex, resource consuming and thus less energy efficient than fixed-point ones.

## 7  Results

The considered system was evaluated twofold – during training of the network and object tracking. In the first case two metrics are considered – the value of

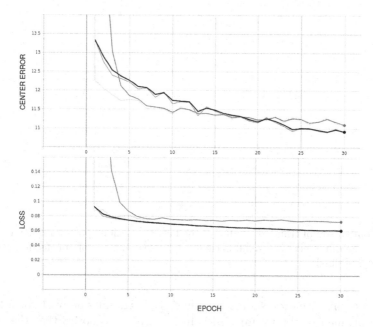

**Fig. 3.** The training progress for the baseline network with floating point precision after 30 epochs. The upper plot shows the centre error, while the lower the loss function. The blue line represents training, while the orange one validation (Color figure online)

the loss function and the average centre error (the difference in pixels between the object's location returned by the network and the ground truth).

In Fig. 3 an exemplar training progress for the baseline floating point network is shown. Due to the small differences between the tested networks, the training results are presented jointly in the Fig. 4 instead of presenting each training curve separately. As expected, the networks with lower precision have greater centre error and loss value (thus lay in upper right corner), on contrary to those with higher precision. The INT16 network slightly outperforms other ones, having a lower centre error. This situation is caused by a different quantisation in the last layer.

The tracking results are presented in Table 3. The tracker was evaluated on a dataset constructed from TempleColor, VOT14, VOT16 set (without OTB sequences – same as in the original paper). The tracking precision is calculated as the ratio of the number of frames with centre error below the predetermined threshold (in our case equal to 20) in relation to the length of the given sequence. The analysis of the average precision values allows to draw interesting conclusions, while comparing the quantised versions to baseline floating point one. First of all, there is a precision growth with increasing quantisation (excluding the INT16 network) with binary network achieving the precision of 44.22 on the top. A similar tendency can be observed using the IOU (Intersection Over Union) metric. The obtained results are not as good as in the original paper,

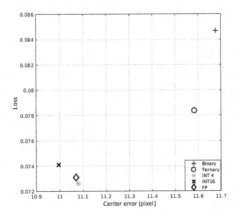

**Fig. 4.** Comparison of training results for the best evaluation epoch for different quantisations. The differences are small because of the low number of epochs, however the trend is maintained

nevertheless this situation should be solved while training network for a longer time (as planned in the future work). It is essential to notice that the binary tracker is relatively better than the one using floating point computations. However, if we acknowledge that neural networks are generally redundant structures, this behaviour is not so astounding, as by quantising weights, the overfitting is prevented. A similar phenomenon was observed in [14] or [20].

**Table 3.** Average tracking precision and average IOU for trackers using quantised Siamese Networks

| Network | Precision [%] | IOU [%] |
|---------|---------------|---------|
| FP | 38.58 | 28.42 |
| INT16[a] | 42.87 | 31.24 |
| INT4 | 36.73 | 27.83 |
| TERNARY | 44.35 | 33.40 |
| BINARY | 44.22 | 32.59 |

[a]Quantisation of the last layer for network INT16 is lower than in other scenarios (INT16 vs INT8).

One experiment was performed with lower output layer quantisation – the network INT16 is using uniform integer quantisation with 16 bits for all convolutional layers. The results show that the precision reduction in the last layer has greater influence on network.

**Inference Profits.** Regardless the lower memory requirements, such quantisations could have impact on the inference time. Given ternary or binary networks, the computational complexity is reduced via replacing standard MAC (Multiply and Accumulate) operations with their binary equivalents. Such operations, if properly implemented, lead to lower execution time. In the case of FPGA implementation it also results in lower resources usage.

## 8    Summary

The presented research explores the possibilities of reducing the computational and memory complexity of a Siamese Neural Network tracker. Two experiments were performed – one during network training and evaluation and the second during tracking. In the first case the quantisation had a negative, as expected, impact on the loss function and centre error. It also turned out that it allowed to reduce the network's tendency to overfit. In the second case, reducing the precision solely in the hidden and slightly in other layers, that is minimising the size of the network in MB, resulted in the increase of the tracker performance. This trend should be further investigated and confirmed on a bigger dataset. The later property is extremely important when aiming for FPGA implementation and enabling highly accurate real-time tracking for a UHD video stream, as a binary network requires more than 10 times less memory than a floating point one.

For future work we plan to perform additional tests using greater number of training data for networks with ternary and binary weights, as the obtained results reveal their tracking potential. For these networks the pruning experiments shall also be executed, as they could allow for even better optimisation. Moreover, more careful analysis on quantisation of input and output, as well as activation layers is needed since such reduction would enable even less memory complexity. Eventually, the final version of the tracker has to be tested on the sequences from the newest VOT challenges for better comparison with state-of-the-art approaches. Finally, the solution should be implemented on a FPGA device.

**Acknowledgements.** The work presented in this paper was supported by the Faculty of Electrical Engineering, Automatics, Computer Science and Biomedical Engineering Dean grant – project number 16.16.120.773 (first author) and the National Science Centre project no. 2016/23/D/ST6/01389 entitled "The development of computing resources organization in latest generation of heterogeneous reconfigurable devices enabling real-time processing of UHD/4K video stream".

## References

1. Held, D., Thrun, S., Savarese, S.: Learning to track at 100 FPS with deep regression networks. In: Leibe, B., Matas, J., Sebe, N., Welling, M. (eds.) ECCV 2016. LNCS, vol. 9905, pp. 749–765. Springer, Cham (2016). https://doi.org/10.1007/978-3-319-46448-0_45

2. Bertinetto, L., Valmadre, J., Henriques, J.F., Vedaldi, A., Torr, P.H.S.: Fully-convolutional Siamese networks for object tracking. In: Hua, G., Jégou, H. (eds.) ECCV 2016. LNCS, vol. 9914, pp. 850–865. Springer, Cham (2016). https://doi.org/10.1007/978-3-319-48881-3_56

3. Guo, Q., Feng, W., Zhou, C., Huang, R., Wan, L., Wang, S.: Learning dynamic Siamese network for visual object tracking. In: 2017 IEEE International Conference on Computer Vision (ICCV), Venice, pp. 1781–1789 (2017)

4. Kristan, M., et al.: The seventh visual object tracking VOT2019 challenge results. In: The IEEE International Conference on Computer Vision (ICCV) Workshops (2019)

5. Valmadre, J., Bertinetto, L., Henriques, J., Vedaldi, A., Torr, P.H.S.: End-to-end representation learning for correlation filter based tracking. In: The IEEE Conference on Computer Vision and Pattern Recognition (CVPR) (2017)

6. Chen, Z., Zhong, B., Li, G., Zhang, S., Ji, R.: Siamese box adaptive network for visual tracking. Accepted to CVPR 2020

7. Courbariaux, M., Hubara, I., Soudry, D., El-Yaniv, R., Bengio, Y.: Binarized neural networks: training deep neural networks with weights and activations constrained to +1 or −1 (2016)

8. Deng, L., Jiao, P., Pei, J., Wua, Z., Li, G.: GXNOR-Net: training deep neural networks with ternary weights and activations without full-precision memory under a unified discretization framework. Neural Netw. **100**, 48–58 (2018)

9. Wu, J., Leng, C., Wang, Y., Hu, Q., Cheng, J.: Quantized convolutional neural networks for mobile devices. In: IEEE Conference on Computer Vision and Pattern Recognition (CVPR) (2016)

10. Han, S., Mao, H., Dally, W.J.: GXNOR-Net: deep compression: compressing deep neural networks with pruning, trained quantization and huffman coding. In: ICLR (2016)

11. Blalock, D., Ortiz, J.J.G., Frankle, J., Guttag, J.: What is the state of neural network pruning? In: Proceedings of Machine Learning and Systems 2020 (MLSys) (2020)

12. Ma, N., Zhang, X., Zheng, H.-T., Sun, J.: ShuffleNet V2: practical guidelines for efficient CNN architecture design. In: Ferrari, V., Hebert, M., Sminchisescu, C., Weiss, Y. (eds.) Computer Vision – ECCV 2018. LNCS, vol. 11218, pp. 122–138. Springer, Cham (2018). https://doi.org/10.1007/978-3-030-01264-9_8

13. Cao, Y., Ji, H., Zhang, W., Shirani, S.: Extremely tiny Siamese networks with multi-level fusions for visual object tracking. In: 22th International Conference on Information Fusion (FUSION) (2019)

14. Zhang, B., Li, X., Han, J., Zeng, Z.: MiniTracker: a lightweight CNN-based system for visual object tracking on embedded device. In: 2018 IEEE 23rd International Conference on Digital Signal Processing (DSP) (2018)

15. Baldi, P., Chauvin, Y.: Neural networks for fingerprint recognition. Neural Comput. **5**(3), 402–418 (1993)

16. Bromley, J., Guyon, I., LeCun, Y., Säckinger, E., Shah, R.: Signature verification using a Siamese time delay neural network. In: Advances in Neural Information Processing Systems 6, [7th NIPS Conference, Denver, Colorado, USA, 1993], pp. 737–744 (1993)

17. Pflugfelder, R.P.: Siamese learning visual tracking: a survey. CoRR, abs/1707.00569 (2017)

18. FINN. https://xilinx.github.io/finn/. Accessed 27 May 2020

19. Yann, L., Denker, J.S., Solla, S.A.: Optimal brain damage. In: Advances in Neural Information Processing Systems, pp. 598–605 (1990)

20. Han, S., Pool, J., Tran, J., Dally, W.J.: Learning both weights and connections for efficient neural networks. In: Proceedings of the 28th International Conference on Neural Information Processing Systems, NIPS 2015, pp. 1135–1143 (2015)
21. Krizhevsky, A., Sutskever, I., Hinton, G.E.: ImageNet classification with deep convolutional neural networks. In: The ACM, vol. 60, pp. 84–90 (2017)

# Nuclei Detection with Local Threshold Processing in DAB&H Stained Breast Cancer Biopsy Images

Lukasz Roszkowiak$^{(\boxtimes)}$ (ID), Jakub Zak (ID), Krzysztof Siemion (ID),
Dorota Pijanowska (ID), and Anna Korzynska (ID)

Nalecz Institute of Biocybernetics and Biomedical Engineering,
Polish Academy of Sciences, Warsaw, Poland
lroszkowiak@ibib.waw.pl
http://ibib.waw.pl/en/

**Abstract.** Histopathological sections allow pathologists to evaluate a wide range of specimens, including breast cancer, obtained from biopsies and surgical procedures. The accuracy of the employed automated cell detection technique is critical in obtaining efficient diagnostic performance. In this paper we investigate 18 different adaptive threshold methods based on various approaches. We validate the methods on a set of histopathological images of breast cancer, where immunohistochemical staining of FOXP3 was performed with 3,3'-diaminobenzidine and hematoxylin. The thresholding is performed on monochromatic images derived from original images: separate channels of Red-Green-Blue and Hue-Saturation-Value, layers of results of color deconvolution, 'brown' channel, 'blue-ratio' layer. The main objective of the evaluation is to determine if the detected objects obtained by the tested methods of thresholding are consistent with the manually labeled ones. The performance is evaluated using precision, sensitivity and F1 score measures. It appears that satisfactory results were achieved only by 6 methods. It was found that *bradley* method is the best performing method for nuclei detection in this type of stained tissue samples. It has best sensitivity value for images after color deconvolution and Value layer (of Hue-Saturation-Value color space), 0.970 and 0.975 respectively. As a result, we recommend a most efficient local threshold technique in the case of nuclei detection in digitized immunohistochemically stained tissue sections. This initial detection of objects followed by texture, size and shape analysis will give a collection of cells' nuclei to perform further accurate segmentation. The proposed detection method will be used in a framework focused on computer-aided diagnosis.

**Keywords:** Histopathology · Digital pathology · Biomedical engineering · Breast cancer

We acknowledge the financial support of the Polish National Science Center grant, PRELUDIUM, 2013/11/N/ST7/02797.

L. J. Chmielewski et al. (Eds.): ICCVG 2020, LNCS 12334, pp. 164–175, 2020.
https://doi.org/10.1007/978-3-030-59006-2_15

# 1   Introduction

The histopathological sections allow pathologists to evaluate a wide range of specimens obtained from biopsies and surgical procedures. Special stains, such as immunohistochemical (IHC) stains, can be employed to augment the diagnosis. The tissue section examination can be performed by an experienced pathologist directly via microscope or using digital images of the samples. The human direct evaluation is irreproducible, time-consuming as well as prone to intra- and interobserver errors. Computer-aided image analysis techniques can be applied to further improve the microscopic analysis.

The processing of the digitized image of stained tissue sample is a complex problem, because of the huge variability in shape, size and color of objects of interest and in the general architecture of the tissue samples. Sections may be sparsely or densely packed by various objects more or less similar to the diagnostically important cells' nuclei. Moreover, some image features hinder the segmentation process, for example: a presence of spurious stain deposits in cell types other than intended (stromal, scar, lymphocytes); very dark parts of blue stained nuclei; partly blurred nuclei border with the color rim caused by the chromatic aberration; color noise. The presence of staining variations and nonuniformity of the background can also obstruct accurate segmentation of the nuclei in histopathological images.

In typical approach to computerized quantitative histopathological image analysis, the detection of cell nuclei is the first major step that can lead to cell segmentation. The accuracy of the employed automated cell detection technique is critical in obtaining good and efficient diagnostic performance. It could lead to identifying the structure or pattern in cells localization based on their detection, as well as features such as density, can be used to support assessing not only cancer grades but also for predicting treatment effectiveness [4,27].

All features of images and objects of interest as well as our prior experience suggest that adaptive threshold methods of segmentation should allow to separate objects or their fragments from digital images of thin tissue slices stained immunohistochemically [18,24].

In this study we compare different methods of thresholding and recommend a most efficient technique in the case of nuclei detection in histopathological images immunohistochemically stained with 3,3'-diaminobenzidine and hematoxylin (DAB&H). We validate the methods of detection on synthetic and manually labeled set of images of breast cancer and axillary lymph nodes tissue. We apply the classical image processing approach, as the proposed detection method will be incorporated into a multipart framework focused on computer-aided diagnosis in the field of histopathology.

## 1.1   Related Works

Many different techniques have been proposed for detection and segmentation of cells' nuclei in digital images of tissue sections. The threshold-based approaches, both global and adaptive, are the most widely used, as it is a simple technique and has low computational complexity. It is particularly important in the case of histopathological images which are large-area digital images. The review [14] shows that threshold-based approach is used in majority for images with hematoxylin and eosin stained tissue samples. Thresholding is generally applied to monochromatic images [6], therefore various threshold-based methods are preceded by extraction [10], deconvolution [31] or combination of color information [2] from image. Various color models are utilized: from Red-Green-Blue (RGB) and process of color signal deconvolution [31], to conversion to other color spaces such as Hue-Saturation-Value (HSV) [19] or CIELab [10].

Deep learning and convolutional neural networks are gradually becoming the state-of-the-art methods in case of image processing with uncertainty and variability of features in objects. Although, the application of deep learning to histopathology is relatively new [20]. Most already published work has focused on the detection of mitotic figures or classification of tissue. The classification of various types of cancer tissue samples can be solved now with deep neural networks. However, the majority of investigations treats most popular hematoxylin and eosin staining, while IHC stained related studies are sparse. Additionally, the detection of separate cells is still a challenging problem even for deep learning algorithms [1]. Furthermore, based on our contact with clinicians, we encountered a strong distrust towards neural network models because of their opaqueness. Based on our co-operation with medical practitioners, whom ultimately will be the end-users of planned framework, we decided to apply classical image processing methods. This will provide clinicians with clearly interpretable algorithm, that is more supportive in everyday practice.

## 2   Materials and Methods

### 2.1   Real Data

The images of breast cancer and axillary lymph nodes tissue used for the validation of the experiments were obtained from the Hospital de Tortosa Verge de la Cinta (Spain). Tissue microarrays were formed by extracting small cylinders of tissue from histological sections and arranging them in an array on a paraffin block so that multiple samples could be processed simultaneously. The prepared sections were immunohistochemically stained with DAB and contra-stained with hematoxylin using an automated stainer. The digital images were captured at 40x magnification on automated whole slide scanning system, and then split into images of separate cases.

All images show the brown end products for the immunopositive cells' nuclei among blue color nuclei for the immunonegative cells. These images differ in color ranges, pattern of objects (nuclei) distribution as well as in local and global contrast and brightness. The inside of brown objects is visible as almost homogeneous, with smooth and slightly visible texture, while the inside of blue objects seems to be filled mostly with fuzzy texture.

To test the methods of segmentation 13 regions of interest of size $1000 \times 1000$ pixels were extracted for quantitative evaluation. The regions of interest were randomly selected from the dataset, by random draw of coordinates that became the upper-left corner of the image. The quality of the image set is typical for this type of biological data. The set consists of images with different degrees of complexity, tissue presented in the images varies from very dark to light and with sparse and compact architecture of cells, containing a total of 7557 cells. For evaluation purposes, ground truth templates were generated by manually annotating the selected regions of interest. The annotations were made by two experts using a free and open-source graphics editor—GIMP. Each expert marked the immunopositive and immunonegative nuclei location with an indicator in separate layers. The layers of marked nuclei were merged into one ground truth template. Example of such annotation is presented in Fig. 1.

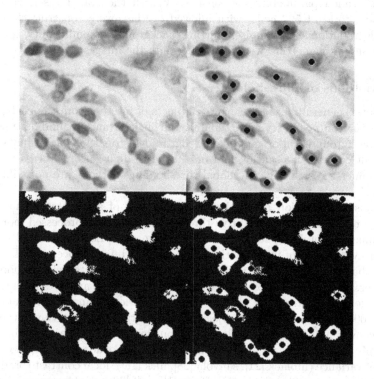

**Fig. 1.** (top-left) Original RGB image. (top-right) Manual annotations of marked nuclei on original image. (bottom-left) Results of nuclei detection by segmentation algorithm. (bottom-right) Results with annotated nuclei location.

Although the images are 3-channel RGB images, we preprocess them to get monochromatic images containing portion of color information:

- The blue channel of RGB.
- The results of color deconvolution ('deconv_DAB', 'deconv_H').
- The Value layer of HSV color representation.

- The 'brown channel' [30], based on values of RGB, calculated as $B-0.3(R+G)$
- The 'blue-ratio layer' [13] defined as $((100*B)/(1+R+G))*(256/(1+B+R+G))$

Then we apply the segmentation algorithms to each monochromatic image separately.

## 2.2  Semi-real Data

The semi-real dataset allows the comparison of segmentation results with the fixed exact position of objects of interest. Artificial images based on the acquired images were generated in the manner described in previous experiments [18], but with tuned parameters for consistency with the real dataset. 20 images containing total of 592 cells, with allowed overlap, were constructed. Typical image degradation such as noise, blurring, background nonuniformity, chromatic aberration and vignetting were introduced into the images.

## 2.3  Methods of Thresholding

Detection is generally used to separate objects of interest from the background. In this situation we are trying to separate the nuclei from surrounding tissue and empty background. Unfortunately the digitized histopathological samples tend to present contrast fluctuation and noise, due to digitization process by whole slide scanners. For this reason global thresholding methods are not sufficiently efficient.

In this investigation all implemented methods of segmentation have been modified to perform local adaptive thresholding on all preprocessed monochromatic images containing separated color information. Local threshold is calculated at every point of image with sliding window image processing. Threshold value is based on the intensity of the analysed pixel and its neighborhood [29]. Moreover, we use simple post-processing to remove small artifacts from the output image. We discriminate objects consisting of less than 50 pixels and more than 10,000 pixels. In addition, objects with solidity (the proportion of the pixels in the convex hull that are also in the region) lower than 0.4 are discriminated.

In this paper we investigated 18 different methods of adaptive threshold based on: local variance (niblack [21], sauvola [28], nick [16]), local contrast (bradley [7], white [32], bernsen [3], yasuda [34], feng [11], wolf [33]), center-surround scheme (palumbo [23]), entropy (entropy [9], maxent [12], kapur [15], mce [8]), clustering (otsu [22]), minimum error (kittler [17]), and fuzzy clustering (fuzzy [26], fuzzcc [5]). The parameters of threshold methods and the size of pixel neighbourhood for local processing were optimized experimentally to achieve best performance. All tested thresholding algorithms were implemented in MATLAB.

## 2.4  Evaluation

The main objective of the evaluation is to determine if the detected objects obtained by the tested methods of thresholding are consistent with the manually labeled ones. To evaluate and to compare the methods with each other two measures were used, namely the precision and the sensitivity of object detection. Precision is the ratio of true positives (TP) to the number of detected objects ($precision = TP/(TP + FP)$). Sensitivity shows what proportion of the objects of interest is found ($sensitivity = TP/(TP + FN)$). In case of both measures, values closer to 1 imply better detection.

Values of these measures are computed as follows. Since the nuclei detection results are provided with a binary image, the set of detected objects is matched with the set of ground truth objects. The centroid of every found object is calculated. A segmented nuclei region is counted as matched if its centroid is localized within an area of the manually labeled nucleus location. True Positive (TP) is then the number of found objects that have a matching ground truth object. False Negative (FN) is the number of ground truth objects without matching object, while False Positive (FP) is the number of detected objects that have no matching ground truth object.

Furthermore, we based our evaluation on F1 score which is calculated as harmonic mean of the sensitivity and precision ($F1score = 2 * (perecision * sensitivity)/(perecision + sensitivity)$). All of these do not measure morphological accuracy, but they show nuclei detection performance.

## 3   Results

We compare the algorithm performance on microscopic images by comparison of the results of nuclei detection.

The thresholding is performed on 78 monochromatic images derived from 13 original RGB images: blue of RGB, two layers of results of color deconvolution, Value of HSV, brown channel, blue-ratio layer. Each image is segmented by 18 local thresholding algorithms. The mean sensitivity and precision measures as well as F1 score for each method were compared to distinguish the best one. Object detection comparison based on F1 score for semi-real data is presented in Fig. 2a and for real data is presented in Fig. 2b.

It appears that satisfactory results were achieved only by following 6 methods: *bradley* (described further in Sect. 4), *feng*, *fuzzycc*, *mce*, *niblack*, *nick* when performed using following types of images: Value of HSV, blue-ratio layer, and H layer separated by color deconvolution. Thus, for clarity, only the results of the best 6 methods are further discussed in this paper, and they are presented in Table 1. The majority of remaining methods achieved very poor results.

## 4   Discussion

The accuracy of the employed automated nuclei detection technique is critical in obtaining good and efficient diagnostic performance of the expected computer-aided diagnosis multipart framework. Based on our previous experience with immunopositive cell nuclei detection, our current approach assumes that it is better to detect (and then segment) both immunopositive and immunonegative

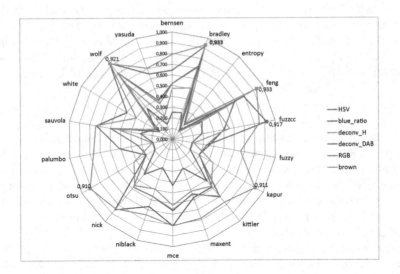

(a) Results on semi-real dataset; the best performance value for the method (above 0.9) displayed in the chart.

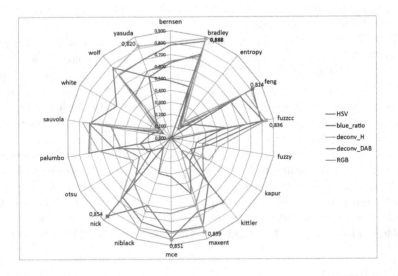

(b) Results on real dataset; the best performance value for the method (above 0.8) displayed in the chart.

**Fig. 2.** F1 score performance of all tested detection algorithms.

**Table 1.** Nuclei detection results. Object detection comparison based on mean sensitivity, mean precision, and F1 score measures. Methods sorted by best sensitivity.

| | | Semi-real dataset | | | | Real dataset | | |
|---|---|---|---|---|---|---|---|---|
| | | Sensitivity | Precision | F1 score | | Sensitivity | Precision | F1 score |
| blue_ratio | nick | 0.970 | 0.767 | 0.857 | nick | 0.976 | 0.760 | 0.854 |
| | feng | 0.960 | 0.672 | 0.791 | mce | 0.870 | 0.813 | 0.841 |
| | mce | 0.954 | 0.701 | 0.808 | fuzzcc | 0.821 | 0.801 | 0.810 |
| | bradley | 0.949 | 0.894 | 0.920 | bradley | 0.775 | 0.721 | 0.747 |
| | fuzzcc | 0.923 | 0.910 | 0.917 | feng | 0.649 | 0.738 | 0.691 |
| | niblack | 0.647 | 0.689 | 0.667 | niblack | 0.493 | 0.770 | 0.601 |
| deconv_H | feng | 0.972 | 0.897 | 0.933 | bradley | 0.970 | 0.819 | 0.888 |
| | nick | 0.972 | 0.748 | 0.846 | mce | 0.916 | 0.795 | 0.851 |
| | bradley | 0.938 | 0.927 | 0.933 | niblack | 0.855 | 0.782 | 0.817 |
| | mce | 0.931 | 0.713 | 0.808 | fuzzcc | 0.787 | 0.893 | 0.836 |
| | niblack | 0.900 | 0.640 | 0.748 | feng | 0.758 | 0.730 | 0.743 |
| | fuzzcc | 0.891 | 0.803 | 0.845 | nick | 0.691 | 0.670 | 0.680 |
| HSV | nick | 0.945 | 0.421 | 0.582 | bradley | 0.975 | 0.774 | 0.863 |
| | feng | 0.899 | 0.558 | 0.689 | feng | 0.929 | 0.725 | 0.814 |
| | mce | 0.888 | 0.535 | 0.668 | mce | 0.908 | 0.694 | 0.787 |
| | bradley | 0.857 | 0.623 | 0.721 | nick | 0.839 | 0.530 | 0.650 |
| | niblack | 0.747 | 0.454 | 0.565 | niblack | 0.828 | 0.682 | 0.748 |
| | fuzzcc | 0.555 | 0.453 | 0.499 | fuzzcc | 0.761 | 0.827 | 0.793 |

cell nuclei simultaneously and then classify selected objects into those two categories or remove them from consideration as insignificant cell types. The main reason for that is a lack of color intensity cut-offs established by pathologists for this type of classification. This problem is especially visible in experts' evaluation where one object may be assigned to different group by different experts. To get around this problem the ground truth and proposed methods in this investigation do not differentiate immunopositive and immunonegative nuclei.

It was found that *bradley* method is the best performing method in the case of nuclei detection in DAB&H stained tissue samples. It has best sensitivity value for images of *deconv_H* (hematoxylin intensity layer extracted by color deconvolution) and Value layer (of HSV), 0.970 and 0.975 respectively. It has also satisfactory values of precision, 0.819 and 0.774 respectively. The great advantage of *bradley* method is only one parameter (T) that has to be set (apart from sliding window size). The concept of the algorithm is that every pixel is considered as background if its brightness is T percent lower than the average brightness of surrounding pixels in the window of the specified size, otherwise it is considered as object of interest. The parameter T was experimentally set to 5 (except for blue-ratio, where it was set to 1), while size of window was 151 (pixels), for scan at 40x magnification. The parameter T is a relative cut-off proportion of pixel intensities in the image (or sliding window), thus the method works better than other methods (*feng, niblack, nick*) with parameter values fixed by user.

It is especially apparent with *feng* method. This method locally calculates standard deviation, that affects the threshold value, and uses two local windows, one contained within the other. The main drawback is its susceptibility to the empirically determined parameter values. Also, it is computationally inefficient with calculation of two sliding windows.

Moreover, *niblack* method adapts the threshold according to the local mean and standard deviation. The results generally suffer from background noise. *Nick* method is an improvement on the *niblack* method. It works well even with images that have very low intensity variations and it minimizes the background noise, but results in low precision in this type of images.

Additionally, *fuzzycc* implements the fuzzy c-means clustering where the clustering criterion used to aggregate subsets is a generalized least-squares objective function. Unfortunately, it lacks the necessary sensitivity to be utilized.

The *mce* method utilizes a non-metric measure, the cross-entropy, to determine the optimum threshold. It is noteworthy as it has no input parameters apart from sliding window size. As a consequence it may pose even more adaptive relevance while processing images, such as stained tissue sections, with high color intensity variability. The sliding window size should be modified accordingly to magnification of the analysed images (151 pixels for 40x in our experiments).

Furthermore, the relation of artificial and real image datasets results is worth mentioning. In most cases we observed slightly lower values for real dataset results in comparison to artificial dataset but with similar overall tendency, which was expected. However, for Value layer of HSV we observe the reversed situation, as could be seen in Table 1. It appears that while processing Value layer (with pixel intensities separated from those providing color information) the artificial images seemed more difficult to process than their real counterparts. Regardless of the fact that introduced background nonuniformity in artificial images is generated according to statistical model based on the set of experimental data, the results indicate that in Value layer of HSV it causes increased similarity between background and objects of interest.

The proposed detection method implements classical approach to image processing by utilizing *bradley* adaptive threshold algorithm. Although deep learning is gradually taking over the image processing in general, it's main drawback is the need for vast amount of labeled data. While there is still lack of IHC datasets available online, the classical image processing approach to this type of images is still a valid and potent technique. Hence, the computer-aided diagnosis multipart framework we are working on may be more appealing for medical practitioners.

## 5   Conclusions and Future Work

In this study we tackle a complex problem of segmentation of the digitized image of stained tissue samples. We apply the classical image processing approach, and investigate various adaptive threshold methods. Conclusively, the *bradley* method as the most efficient detection method, will be a fundamental component in a multipart framework for immunohistochemically stained histopathological images. The throughout validation performed on synthetic and manually labeled set of images of breast cancer and axillary lymph nodes assures about relevance of achieved results.

In this investigation we focus on detection of the nuclei in images of stained tissue. Thus we make the assessment based only on the number of objects, consequently the used measures do not characterize morphological accuracy. This initial detection of objects followed by texture, size and shape analysis will give a collection of cells' nuclei to perform further accurate segmentation. To obtain better segmentation results, we consider using active contour or k-means method to adjust the nuclei boundary and then perform morphology accuracy evaluation [25].

Since, the proposed detection methodology will be merely first stage of processing in expected computer-aided diagnosis multipart framework, several other issues remain to be addressed. For example, the problem of overlapping and clustered nuclei is still a major challenge in the field of nuclei segmentation. We plan to investigate a variety of methods to achieve best results.

Moreover, the detection results in objects representing both immunopositive and immunonegative nuclei. To distinguish between those two classes we plan to apply postprocessing step with discrimination criteria based on machine learning. We plan to use neural network with feature extraction to achieve classification similar to human-expert level.

**Funding.** We acknowledge the financial support of the Polish National Science Center grant, PRELUDIUM, 2013/11/N/ST7/02797. The funders had no role in study design, data collection and analysis, decision to publish, or preparation of the manuscript.

**Acknowledgment.** We would like to thank Molecular Biology and Research Section, Hospital de Tortosa Verge de la Cinta, Institut dInvestigaci Sanitria Pere Virgili (IISPV), URV, Spain and Pathology Department of the same hospital for their cooperation and generous sharing of samples.

# References

1. Aresta, G., et al.: BACH: grand challenge on breast cancer histology images. Med. Image Anal. **56**, 122–139 (2019). https://doi.org/10.1016/j.media.2019.05.010
2. Basavanhally, A., et al.: Computerized image-based detection and grading of lymphocytic infiltration in HER2+ breast cancer histopathology. IEEE Trans. Biomed. Eng. **57**(3), 642–653 (2010). https://doi.org/10.1109/tbme.2009.2035305
3. Bernsen, J.: Dynamic thresholding of gray-level images. In: ICPR 1986: International Conference on Pattern Recognition, pp. 1251–1255 (1986)
4. Beyer, M.: Regulatory T cells in cancer. Blood **108**(3), 804–811 (2006). https://doi.org/10.1182/blood-2006-02-002774

5. Bezdek, J.C., Ehrlich, R., Full, W.: FCM: the fuzzy c-means clustering algorithm. Comput. Geosci. **10**(2–3), 191–203 (1984). https://doi.org/10.1016/0098-3004(84)90020-7

6. Bogachev, M.I., Volkov, V.Y., Kolaev, G., Chernova, L., Vishnyakov, I., Kayumov, A.: Selection and quantification of objects in microscopic images: from multicriteria to multi-threshold analysis. BioNanoScience **9**(1), 59–65 (2018). https://doi.org/10.1007/s12668-018-0588-2

7. Bradley, D., Roth, G.: Adaptive thresholding using the integral image. J. Graph. Tools **12**(2), 13–21 (2007). https://doi.org/10.1080/2151237X.2007.10129236

8. Brink, A., Pendock, N.: Minimum cross-entropy threshold selection. Pattern Recognit. **29**(1), 179–188 (1996). https://doi.org/10.1016/0031-3203(95)00066-6

9. Davies, E.R.: Machine vision: theory, algorithms, practicalities. Elsevier (2004). https://www.elsevier.com/books/machine-vision/davies/978-0-12-206093-9

10. Dundar, M.M., et al.: Computerized classification of intraductal breast lesions using histopathological images. IEEE Trans. Biomed. Eng. **58**(7), 1977–1984 (2011). https://doi.org/10.1109/tbme.2011.2110648

11. Feng, M.L., Tan, Y.P.: Contrast adaptive binarization of low quality document images. IEICE Electron. Exp. **1**(16), 501–506 (2004). https://doi.org/10.1587/elex.1.501

12. Gargouri, F., et al.: Automatic localization methodology dedicated to brain tumors based on ICP matching by using axial MRI symmetry. In: 2014 1st International Conference on Advanced Technologies for Signal and Image Processing (ATSIP), pp. 209–213. IEEE (2014). https://doi.org/10.1109/atsip.2014.6834608

13. Irshad, H.: Automated mitosis detection in histopathology using morphological and multi-channel statistics features. J. Pathol. Inform. **4**(1), 10 (2013). https://doi.org/10.4103/2153-3539.112695

14. Irshad, H., et al.: Methods for nuclei detection, segmentation, and classification in digital histopathology: a review—current status and future potential. IEEE Rev. Biomed. Eng. **7**, 97–114 (2014). https://doi.org/10.1109/rbme.2013.2295804

15. Kapur, J., et al.: A new method for gray-level picture thresholding using the entropy of the histogram. Comput. Vis. Graph. Image Process. **29**(3), 273–285 (1985). https://doi.org/10.1016/0734-189x(85)90125-2

16. Khurshid, K., et al.: Comparison of niblack inspired binarization methods for ancient documents. In: Berkner, K., Likforman-Sulem, L. (eds.) Document Recognition and Retrieval XVI, p. 72470U. SPIE (2009). https://doi.org/10.1117/12.805827

17. Kittler, J., Illingworth, J.: Minimum error thresholding. Pattern Recognit. **19**(1), 41–47 (1986). https://doi.org/10.1016/0031-3203(86)90030-0

18. Korzynska, A., et al.: Validation of various adaptive threshold methods of segmentation applied to follicular lymphoma digital images stained with 3,3'-diaminobenzidine&haematoxylin. Diagnost. Pathol. **8**(1), 1–21 (2013). https://doi.org/10.1186/1746-1596-8-48

19. Kuse, M., Sharma, T., Gupta, S.: A classification scheme for lymphocyte segmentation in H&E stained histology images. In: Ünay, D., Çataltepe, Z., Aksoy, S. (eds.) ICPR 2010. LNCS, vol. 6388, pp. 235–243. Springer, Heidelberg (2010). https://doi.org/10.1007/978-3-642-17711-8_24

20. Litjens, G., et al.: Deep learning as a tool for increased accuracy and efficiency of histopathological diagnosis. Sci. Rep. **6**(1) (2016). https://doi.org/10.1038/srep26286

21. Niblack, W.: An Introduction to Image Processing. Prentice-Hall International, Englewood Cliffs(1986). http://books.google.pl/books?id=XOxRAAAAMAAJ

22. Otsu, N.: A threshold selection method from gray-level histograms. IEEE Trans. Syst. Man Cybern. **9**(1), 62–66 (1979)

23. Palumbo, P.W., et al.: Document image binarization: Evaluation of algorithms. In: Proceedings of SPIE, Applications of Digital Image Processing IX, vol. 697, no. 278, pp. 278–285 (1986). https://doi.org/10.1117/12.976229

24. Roszkowiak, L., et al.: Short survey: adaptive threshold methods used to segment immunonegative cells from simulated images of follicular lymphoma stained with 3,3'-diaminobenzidine&haematoxylin. In: Proceedings of the 2015 Federated Conference on Computer Science and Information Systems, pp. 291–296. IEEE (2015). https://doi.org/10.15439/2015f263

25. Roszkowiak, L., Korzyńska, A., Siemion, K., Pijanowska, D.: The influence of object refining in digital pathology. In: Choraś, M., Choraś, R.S. (eds.) IP&C 2018. AISC, vol. 892, pp. 55–62. Springer, Cham (2019). https://doi.org/10.1007/978-3-030-03658-4_7

26. Sarkar, S., Paul, S., Burman, R., Das, S., Chaudhuri, S.S.: A fuzzy entropy based multi-level image thresholding using differential evolution. In: Panigrahi, B.K., Suganthan, P.N., Das, S. (eds.) SEMCCO 2014. LNCS, vol. 8947, pp. 386–395. Springer, Cham (2015). https://doi.org/10.1007/978-3-319-20294-5_34

27. Sasada, T., et al.: CD4+CD25+ regulatory T cells in patients with gastrointestinal malignancies. Cancer **98**(5), 1089–1099 (2003). https://doi.org/10.1002/cncr.11618

28. Sauvola, J., Pietikainen, M.: Adaptive document image binarization. Pattern Recogn. **33**(2), 225–236 (2000). https://doi.org/10.1016/S0031-3203(99)00055-2

29. Sezgin, M., Sankur, B.: Survey over image thresholding techniques and quantitative performance evaluation. J. Electron. Imaging **13**(1), 146–168 (2004). https://doi.org/10.1117/1.1631315

30. Tadrous, P.: Digital stain separation for histological images. J. Microsc. **240**(2), 164–172 (2010). https://doi.org/10.1111/j.1365-2818.2010.03390.x

31. Veillard, A., et al.: Cell nuclei extraction from breast cancer histopathologyimages using colour, texture, scale and shape information. Diagnost. Pathol. **8**(Suppl 1), S5 (2013). https://doi.org/10.1186/1746-1596-8-s1-s5

32. White, J.M., Rohrer, G.D.: Image thresholding for optical character recognition and other applications requiring character image extraction. IBM J. Res. Dev. **27**(4), 400–411 (1983). https://doi.org/10.1147/rd.274.0400

33. Wolf, C., Jolion, J.M.: Extraction and recognition of artificial text in multimedia documents. Form. Pattern Anal. Appl. **6**(4), 309–326 (2004). https://doi.org/10.1007/s10044-003-0197-7

34. Yasuda, Y., et al.: Data compression for check processing machines. Proc. IEEE **68**(7), 874–885 (1980). https://doi.org/10.1109/PROC.1980.11753

# Video Footage Highlight Detection in Formula 1 Through Vehicle Recognition with Faster R-CNN Trained on Game Footage

Ruan Spijkerman and Dustin van der Haar$^{(\boxtimes)}$

University of Johannesburg, Kingsway Avenue and University Rds, Auckland Park,
Johannesburg 2092, South Africa
ruan.spijkerman@hotmail.com, dvanderhaar@uj.ac.za

**Abstract.** Formula One, and its accompanying e-sports series, provides viewers with a large selection of camera angles, with the onboard cameras oftentimes providing the most exciting view of events. Through the implementation of three object detection pipelines, namely Haar cascades, Histogram of Oriented Gradient features with a Support Vector Machine, and a Faster Region-based Convolutional Neural Network (Faster R-CNN), we analyse their ability to detect the cars in real-life and virtual onboard footage using training images taken from the official F1 2019 video game. The results of this research concluded that Faster R-CNNs would be best suited for accurate detection of vehicles to identify events such as crashes occurring in real-time. This finding is evident through the precision and recall scores of 97% and 99%, respectively. The speed of detection when using a Haar cascade also makes it an attractive choice in scenarios where precise detection is not important. The Haar cascade achieved the lowest detection time of only 0.14 s per image at the cost of precision (71%). The implementation of HOG features classifier using an SVM was unsuccessful with regards to detection and speed, which took up to 17 s to classify an image. Both the Haar cascade and HOG feature models improved their performance when tested on real-life images (76% and 67% respectively), while the Faster R-CNN showed a slight drop in terms of precision (93%).

**Keywords:** Object detection · Histogram of oriented gradients · Faster R-CNN · Haar cascade

## 1 Introduction

Thirty-five years after the first Formula One race, in 1985, an onboard camera was attached to Francois Hesnault's Renault during the German Grand Prix [1]. This marked the first time that onboard cameras were used with the main purpose of providing viewers with a better entertainment experience. Previously,

© Springer Nature Switzerland AG 2020
L. J. Chmielewski et al. (Eds.): ICCVG 2020, LNCS 12334, pp. 176–187, 2020.
https://doi.org/10.1007/978-3-030-59006-2_16

this technology was reserved for testing, and even during the 1985 race it was only done as Hesnault's car was not in contention for championship points.

In recent years, however, multiple onboard cameras have become a requirement in Formula One as sanctioned by the Federation Internationale De L'Automobile (FIA) [2]. The ability to see a race from the driver's perspective provide benefits to teams, race marshals, and spectators. With twenty cars spread across a circuit during races lasting up to 120 min, entire teams are required to find highlights among all the footage. The research done through this paper attempts to identify an accurate, real-time classification method to help automate the process of finding highlights captured by the onboard cameras. Highlight footage that might be important to a Formula One spectator includes cars being overtaken, cars going off the race track, and cars colliding. This is due to the impact these events have on the result of the race. The object detection algorithms will thus also be tested against footage from a real-life race in order to determine if the performance would be suitable for the detection of the above-mentioned events.

Due to the first few races of the 2020 season being cancelled or postponed, Formula 1 has announced a virtual Grand Prix series. This series included current and former Formula 1 divers as well as professional e-sports drivers [3]. These races are purely for entertainment and aim to spread interest in Formula 1. The involvement of professional racers and the Formula 1 organisers emphasises the validity of e-sports as a form of entertainment with a need for highlight detection. It also shows the real-world similarity of the game as it can serve as an introduction to Formula 1.

The first section of this paper will be a further breakdown of the problem environment, as well as what defines a successful solution. Following this, a literature review will be presented, and a discussion on similar systems that have been developed. Similar systems will include implementations of Haar cascades, Faster R-CNNs, and HOG features with a classifier for vehicle detection. Through these sections, it should become clear why the proposed pipelines were selected. The remaining sections will feature a description of how each pipeline was set up, tested, and draws any results and conclusions from it.

## 2   Background

As Formula 1 has expanded to include both physical and virtual races, the following section will discuss the use of the Formula 1 video game as a suitable research environment. Following that, the literature study will discuss similar systems for each of the computer vision algorithms implemented in this experiment. Similar systems will be described with regards to their goal, advantages, and disadvantages.

### 2.1   Problem Environment

In recent years Formula One has grown to include an official e-sports series as well as the traditional world championship [4]. The E-sports World Championship is played using the yearly release of the official Formula One video game that aims to be as close to the traditional sport as possible. This realism extends to the broadcasting and viewing of the races.

As the game realistically depicts both the cars and circuits, this research will use the F1 2019 game to capture video data. The main advantage of using game footage is the ability to capture the exact data needed to train the proposed models using the in-game director feature. The game is also graphically realistic enough that the results obtained with a model trained on in-game images may be comparable when tested on real-life footage.

The use of audio and video, taken from Formula 1 television broadcasts, with the goal of highlight detection has been proven to be viable [5]. Using Dynamic Bayesian Networks researchers were able to identify events such as the start of the race with high precision (greater than 80%). The method made use of commentators' voice volume, colour and motion, as well as text that superimposed on the video frames from the broadcast. Although generally successful, the method was unable to passing cars with high precision or recall and achieved roughly 30% for these metrics. In a paper on the detection of cars in IndyCar races, it was found that pre-trained Single Shot Detectors had poor performance with regards to detecting race cars [6]. They achieve their goal of comparing mobile networks with regards to race car detection the authors of the 2019 paper labelled their own dataset of IndyCar race cars. They found, however, that there is a lack of data and that it is especially true for anomalous events such as collisions.

## 2.2  Methods

Computer vision pipelines generally consist of preprocessing, feature extraction, and classification. For each of the pipelines considered in this paper, the preprocessing stage will remain largely the same. Preprocessing diminishes the impact of noise in data and reduces the computational resources required for further processing. In the case of object detection, specifically, algorithms such as Haar cascades and HOG feature vectors trained on downscaled images are able to detect objects independent of their scale [7]. Resizing images improves the performance of the feature extraction and classification phases, with a minimal overhead cost. Gaussian Filters are often used to reduce the noise in images, which is especially important when attempting to detect edges and similar features [8]. Applying Gaussian blurring does, however, reduce the contrast of edges which could decrease the accuracy of algorithms used for edge detection. Converting an image to greyscale from RGB or a similar representation reduces the dimensionality of each pixel from 3 values to 1. The use of greyscale images can thus reduce computational complexity while simultaneously achieving higher accuracy in classification problems [9].

Viola and Jones [10] describe a method for real-time, face detection through the use of Haar-like features that can be calculated in constant time regardless of scale once they have been identified. Their implementation of a cascade allowed for more complex features to be tested in only areas where a face is already likely to exist. Haar features under abnormal lighting conditions are often less accurate; however, the use of improved feature descriptors [11] still make Haar cascades an attractive choice.

Yan et al. attempted to detect vehicles of various shapes in complex environments by their shadows [12]. For this detection, the authors extracted the shadows underneath each vehicle along with HOG feature and trained a linear Support Vector Machine (SVM) on the extracted features as well as the HOG features of negative images in order to detect vehicles. The shadows were used to predict where vehicles would be, however, shadows of nearby vehicles could overlap, and the resulting area would not be found to contain expected HOG features.

Another implementation by Xu et al. uses HOG features detects vehicles from aerial images and videos [13]. Their implementation combined it with the Viola-Jones algorithm and SVM and presented a switching strategy to maintain accuracy and performance. This allowed for real-time performance as well as allowing the detection to work without the prior mapping of surveyed roads. The method was not able to reliably detect turning vehicles which greatly reduces its applicability in real-world scenarios.

Regional Convolutional Neural Networks (R-CNN) have been used with great success with regards to object detection. This is largely due to their ability to extract and represent features using their convolving filters [14]. Two-stage detectors such as R-CNN, Fast R-CNN, and Faster R-CNN make use of a CNN applied to the proposed regions for object detection. These regions, or anchors, are generated by a region proposal network (RPN) that predicts the likelihood of a region containing an object and to adjust the anchor.

In a study on vehicle detection using CNN based classifiers, a Faster R-CNN model was modified with the aim of detecting smaller vehicles [15]. This was done by adding an additional scale of $64^2$ whereas the previous smallest scale was $128^2$. The aspect ratios were left unchanged as 1:1, 1:2, and 2:1. The classifier was trained to detect cars only, meaning that vans, buses, and other large vehicles were not detected.

Liu et al. [14] argued that while R-CNN based detectors are the most effective at general object detection, they are ill-suited for small object detection. Their proposed solution included a Backward Feature Enhancement Network and a Spiral Layout Preserving Network. This allowed them to obtain region proposals with a higher recall than Faster R-CNN based detectors which greatly improved the detection of small vehicles. This was, however, done at the expense of real-time performance.

Real-time performance is crucial when detecting vehicles for self-driving cars. For this reason, HybridNet provides a two-stage method with speeds comparable to those of single-stage methods [16]. This is done by obtaining rough bounding boxes in the initial stage and following it with a more comprehensive search in the second stage. This method resulted in a loss of accuracy in order to obtain real-time performance. Robust, real-time detection was also achieved through the combination of Haar-like features and neural networks [17]. However, it was also a binary classification system, which limited the application to only detecting vehicles and not vehicle types, similar to the system designed by Zhang, Wan, and Bian [15, 17].

## 3  Experiment Setup

For the training and testing of the pipelines implemented for this research, a sufficient number of labelled images from game footage were required as a dataset. Real-life images also have to be captured and label. Both types of images were then preprocessed before having their features extracted that would be used for vehicle detection. This section will thus describe the dataset details, followed by the specifics of the preprocessing steps. Lastly, the configuration of the feature extraction and classification steps will be discussed and justified. The objective of the models presented in this paper is to determine the potential of the discussed pipelines for accurate, real-time detection of F1 cars. The pipelines are thus not necessarily directly compared to each other (Fig. 1).

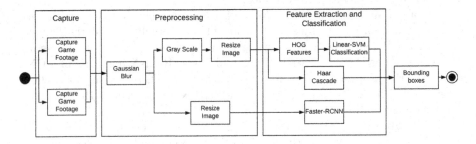

**Fig. 1.** Overview of the complete process showing where pipelines diverge

### 3.1  Dataset Construction

Video footage was captured from the official F1 2019 racing game on PC from the onboard camera perspective that is similar to the onboard camera footage that exists in Formula One today. The video was captured at a framerate of 30 frames per second and a resolution of 720p (1280 × 720 px) using Open Broadcasted Software (OBS). The videos were then split into frames, and unique individual frames were kept.

LabelImg was used to label cars in the images with Extensible Mark-up Language files describing the bounding boxes of the labelled cars. In total, 786 positive images were captured and labelled with a total of 1459 cars labelled in them. A further 479 negative images were captured that contained no cars. Three 426 × 240-pixel areas were then selected from each image to obtain 1437 negative samples.

As discussed earlier, Akkas et al. [6] found the lack of motorsport image datasets to be a limiting factor in their research. As our models are trained on game footage that can be easily generated, it can potentially address this problem if it performs well on real-life footage. In order to test the models' performance on real-life footage, highlight clips were taken from the official Formula 1 Youtube Channel [18]. These videos were downloaded in mp4 format at the same resolution as the in-game footage (720p). In total, over 3000 frames were extracted from the highlight clips. After removing frames that included excessive overlays and graphics, a random subset of 300 frames was selected and manually labelled as was done with the in-game footage (Fig. 2).

(a) Game Footage                           (b) Real-life Footage

**Fig. 2.** A comparison of an image taken from game footage and a real-life image after preprocessing

### 3.2   Preprocessing

The preprocessing step is done to remove noise and reduce the computational cost of feature extraction and classification. The preprocessing set also serves to reduce potential differences between the virtual and real-life footage.

**Gaussian Blur.** The main purpose of the Gaussian Blur method is to reduce noise in the captured images that could be present due to screen tearing or other artefacts present due to the video being captured in-game. A 5 by 5 kernel was used as it provided a satisfactory balance between performance cost and effect. The Gaussian blur preprocessing step was applied to images for each pipeline.

**Conversion to Greyscale.** All images were originally captured with RGB pixel representations. These were then converted to grayscale representations as grayscale images reduce the computational complexity required for further feature extraction. Images were converted to their grayscale representations for both the Haar cascade and HOG feature pipeline, but not for Faster R-CNN detection. This was done as the filters in the convolutional layers could derive features from the colour space [19].

**Resizing.** The resizing of images was a crucial step in minimising the performance cost during feature extraction and classification. The images are sized to maximise performance for the different classifiers within each pipeline. The sizes are therefor different for each pipeline as they are not directly compared. For object detection using the Haar cascade classifier, images were resized to a size of $320 \times 240$ pixels for the tested cascades.

The Support vector machine in the second pipeline was trained on HOG feature representations of images that had a resolution of $426 \times 240$ pixels. Input images were kept at their original resolutions with a $426 \times 240$ frame passing over it to detect cars within that section of the image.

Images were resized to a resolution of $320 \times 180$ pixels for use with the Faster R- CNN pipeline.

### 3.3  Haar Cascade

This pipeline consisted of a cascade that was trained using the dataset described in Sect. 3.1. The classifier was implemented through the OpenCV library, with the following parameters (Table 1):

**Table 1.** Haar cascade detector parameters

| Scale factor | Min neighbours | Min size | Max size |
|---|---|---|---|
| 1.10 | 40 | None | [160, 120] |

The large Min Neighbours parameter was set to help prevent the detection of false positives. Similarly, an increase in the Max Size prevented detection of any objects that were too large to be a vehicle in the resized images.

### 3.4  HOG Features with an SVM Classifier

This pipeline consisted of a Linear Support Vector Machine that was trained on the HOG features of both positive and negative images. The pipeline worked as follows: Firstly, the HOG features were extracted from a set of positive and negative images that had a resolution of $426 \times 240$ pixels. These features with their labels were then used to train an SVM.

Thereafter, test images with a resolution of $1280 \times 720$ pixels were scanned with a moving window of $426 \times 240$ pixels. The HOG features were then extracted from each window and classified using the previously trained SVM. If the window was found to contain a false positive, the false-positive frame would be saved and labelled as a negative image for the purpose of hard negative mining.

The original training set, combined with the false positives detected in the previous step, was used to retrain the SVM to ensure regions similar to the previously detected false positives would be classified as negatives. Finally, detection would also utilise a sliding window, and store the minimum and maximum $x$ and $y$ coordinates of all windows that were classified as containing a car. A Non-Maximum Suppression method is then used to merge overlapping detections and limit the detection of false positives. For this method, an Intersection over Union value of 0.2 was used as it provided a fair ratio between false negatives and false positives.

### 3.5   Faster R-CNN

This pipeline was based on the pre-trained Faster R-CNN Inception v2 model from the Tensorflow object detection model zoo [20]. This model was selected as it provided the best balance between speed and accuracy when compared to the other models available from the Tensorflow object detection model zoo.

The model was further trained using TensorFlow-GPU and the dataset described in Sect. 3.1. The default scale values of 0.25, 0.5, 1.0, and 2.0 were maintained, as well as the default aspect ratios of 1:2, 1:1, and 2:1. The model was trained for 20311 steps using 575 training images and 40 epochs. As seen in the figure below, further training was found to be unnecessary, as the loss rate plateaued at a value of 0.05 (Fig. 3).

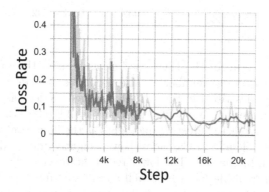

**Fig. 3.** Graph depicting the smoothed loss rate of the trained model over 40 epochs. The background line shows the loss rate at each step

## 4   Results

The results of each classifier are based on their performance on test images containing between 1 and 3 cars each. Classifiers were evaluated based on their precision, recall, F1 score, and detection time. This section will contain the results of the SVM that was trained using the HOG features as well as the results of the complete pipeline as it will help identify potential flaws in the implementation.

## 4.1   Pipeline Metrics

**Table 2.** Results of individual pipelines on game footage with detection time measured in seconds per frame

| Pipeline | Precision | Recall | F1 score | Detection time |
|---|---|---|---|---|
| Haar cascade | 71% | 49% | 58% | 14 s |
| HOG features with SVM | 47% | 62% | 53% | 17 s |
| Faster R-CNN | 97% | 99% | 98% | 1.4 s |

**Table 3.** Results of individual pipelines on the real-life footage. Detection time is omitted as it was invariant

| Pipeline | Precision | Recall | F1 score |
|---|---|---|---|
| Haar cascade | 76% | 32% | 45% |
| HOG features with SVM | 67% | 88% | 76% |
| Faster R-CNN | 93% | 91% | 92% |

Based on the low recall of the Haar cascade as seen in the tables above, the pipeline resulted in a large number of false negatives. The decrease was caused by the parameters described in Sect. 3.3 that aimed to decrease the number of false positives. A larger number of false positives would reduce the effectiveness of the system with regards to detecting highlights as the aim is only to detect when vehicles are present. The Haar Cascades showed slightly improved precision when tested against real-life images. The lower recall, however, indicates that it does not detect cars in real-life footage as well as in-game footage. The parameters discussed in Sect. 3.3 could potentially be adjusted to improve this. However, the aim of this research is, in part to show the performance of a model trained on virtual images applied directly in a real-world environment (Tables 2 and 3).

The Haar cascade was, however, notably faster than the other methods tested and would be applicable in the real-time detection of vehicles. The HOG features performed noticeably worse in this regard due to the inefficient searching using a sliding window over the image. Reducing the area being considered could provide a significant improvement in future implementations.

The SVM used in conjunction with the HOG feature vectors obtained a precision of 83% and a recall of 81%. Comparing the SVM's classification ability on the training images with those obtained using the sliding window suggests the dataset or sliding window caused the loss in precision. Firstly, there is a potential indication that the training dataset images were cropped too tightly around the car and thus did not allow for the correct detection in the testing dataset that included more background around the car. Another problem could be the implementation of the sliding window. The size of this window could be adjusted to more closely represent the size of a vehicle in the test images. Although the speed of the detection was inferior in both cases, the precision and recall improved when tested on the real-life dataset. Further research may investigate the cause of the improved performance, as it stands in contrast to the other pipelines.

The Faster R-CNN was the only pipeline that did not produce a large number of false negatives and clearly outperformed the other pipelines in each aspect other than speed. Potential alternatives such as a single-shot detection and You Only Look Once could provide a faster detection time on lower-end hardware. It should also be noted that the detection time was measured on a PC running TensorFlow-CPU as the other pipelines also made use of CPU processing power. When tested on a system running TensorFlow-GPU and an NVIDIA GTX 1050 GPU the speed increased to 0.56 s per image. The results remained largely the same when tested on real-life images. There was a slight increase in false negatives due to effects such as smoke obscuring visibility, as seen in the figure below. Although these effects are present in-game, they are often not as severe as in the real-life footage (Fig. 4).

**Fig. 4.** Example output of the Haar cascade (a), HOG feature (b), and Faster R-CNN (c and d) pipelines. (a) and (b) clearly show a large number of false positive detections. The last image shows a false negative caused by smoke caused by a car's brakes locking up

## 4.2  Discussion

The results show in this paper is promising for the use of computer vision with regards to highlighting detection in Formula 1 in virtual and traditional forms. Examples of detected highlights might include two detected cars coming close together and then quickly moving apart as a collision event. One car appearing suddenly from the side, then moving to the centre would be an indication of an overtake of the car being followed. While two cars in the centre moving close together and then only a single car being detected could be an indication of an overtake being seen from behind.

Another promising avenue to further investigate is the use of virtual images to train models that can be used in real-world settings. This allows us to create or expand datasets that include specific images that would not usually be easily obtainable. For example, a significant collision in Formula 1 involving multiple cars is a relatively rare occurrence and would be extremely costly to recreate for the purpose of building a dataset.

## 5  Conclusion

Based on the implemented pipelines, it is clear that Faster R-CNNs would be best suited for the real-time detection of highlights such as crashes occurring during Formula One races from onboard footage. Although the low precision of the other classifiers makes them unable to detect two cars touching, they would still be able to identify cars being overtaken from behind from the perspective of the car being overtaken.

Future research could be focused on either improving the performance of the identified classifiers within the presented problem domain or on creating new models for other real-world environments using datasets created in virtual environments. Two potential domains could be aeroplane detection and recognition, or the detection of infantry and military vehicles. The reason for this is the existence of hyper-realistic flight and military simulation games. The use of games could allow us to create datasets of these objects with specific, life-like backgrounds that may improve the detection thereof under similar circumstances. Furthermore, there may be value in researching the fine-tuning of a Faster R-CNN to detect smaller cars following the research identified in Sect. 2 of this paper. Other possible avenues for further experimentation could include improvements to the recall of Haar cascades in this environment as well as improving the sliding window used in the HOG features pipeline.

## References

1. First Onboard Camera Telecast in Formula. https://www.auto123.com/en/news/f1-first-onboard-camera-telecast-in-formula-1/35450/. Accessed 23 Oct 2019
2. Fédération Internationale de l'Automobile: 2020 F1 Technical Regulations, Paris, p. 54 (2019)

3. Formula 1 launches Virtual Grand Prix Series to replace postponed races. https:// www.formula1.com/en/latest/article.formula-1-launches-virtual-grand-Prix- series-to-replace-postponed-races.1znLAbPzBbCQPj1IDMeiOi.html. Accessed 2 Apr 2020
4. F1 Esports Series 2019. https://f1esports.com/. Accessed 23 Oct 2019
5. Petkovic, M., Mihajlovic V., Jonker, W., Djordjevic-Kajan, S.: Multi-modal extrac- tion of highlights from TV Formula 1 programs. In: IEEE International Conference on Multimedia and Expo, pp. 817–820. IEEE, Switzerland (2002)
6. Akkas, S., Maini, S.S., Qiu, J.: A fast video image detection using tensorflow mobile networks for racing cars. In: IEEE International Conference on Big Data, pp. 5667– 5672. IEEE, USA (2019)
7. Mishra, G., Aung, Y.L., Wu, M., Lam, S., Srikanthan, T.: Real-time image resizing hardware accelerator for object detection. In: International Symposium on Elec- tronic System Design, pp. 98–102. IEEE, Singapore (2013)
8. Sarfraz, M., Taimur, A.: Real time object detection and motion. In: Second Inter- national Conference in Visualisation, pp. 235–240. IEEE, Barcelona (2009)
9. Bui, H.M., Lech, M., Cheng, E., Neville, K., Burnett, I.S.: Object recognition using deep convolutional features transformed by a recursive network structure. Access IEEE 4(1), 10059–10066 (2016)
10. Jones, M., Viola, P.: Rapid object detection using a boosted cascade of simple fea- tures. In: Proceedings of Conference on Computer Vision and Pattern Recognition, pp. 511–518. IEEE, USA (2001)
11. Park, K., Hwang, S.: An improved Haar-like feature for efficient object detection. Pattern Recogn. Lett. 42(1), 148–154 (2014)
12. Yan, G., Yu, M., Yu, Y., Fan, L.: Real-time vehicle detection using histograms of oriented gradients and AdaBoost classification. Optik 127(19), 7941–7951 (2016)
13. Xu, Y., Yu, G., Wang, Y., Wu, X., Ma, Y.: A hybrid vehicle detection method based on Viola-Jones and HOG + SVM from UAV images. Sensors 16(8), 7941– 7951 (2019)
14. Liu, W., Liao, S., Hu, W.: Towards accurate tiny vehicle detection in complex scenes. Neurocomputing 347, 22–33 (2019)
15. Zhang, Q., Wan, C., Bian, S.: Research on vehicle object detection method based on convolutional neural network. In: 11th International Symposium on Computational Intelligence and Design, pp. 271–274. IEEE, China (2018)
16. Dai, X.: HybridNet: a fast vehicle detection system for autonomous driving. Sig. Process. Image Commun. 70, 79–88 (2019)
17. Atibi, M., Atouf, I., Boussaa, M., Bennis, A.: Real-time detection of vehicles using the Haar-like features and artificial neuron networks. In: The International Con- ference on Advanced Wireless, Information, and Communication Technologies, pp. 24–31. Procedia Computer Science, Tunisia (2015)
18. Formula 1. https://www.youtube.com/user/Formula1. Accessed 2 Apr 2020
19. Shao, F., Wang, X., Meng, F., Zhu, J., Wang, D., Dai, J.: Improved faster R- CNN traffic sign detection based on a second region of interest and highly possible regions proposal network. Sensors 19(10), 228–256 (2019)
20. Tensorflow detection model zoo. https://github.com/tensorflow/models. Accessed 21 Oct 2019

# Performance Evaluation of Selected 3D Keypoint Detector–Descriptor Combinations

Paula Stancelova[1]([envelope]) [ID], Elena Sikudova[2] [ID], and Zuzana Cernekova[1] [ID]

[1] Faculty of Mathematics, Physics and Informatics,
Comenius University in Bratislava, Bratislava, Slovakia
`paula.budzakova@fmph.uniba.sk, zuzana.cernekova@uniba.sk`
[2] Faculty of Mathematics and Physics, Charles University,
Prague, Czech Republic
`sikudova@cgg.mff.cuni.cz`

**Abstract.** Nowadays, with easily accessible 3D point cloud acquisition tools, the field of point cloud processing gained a lot of attention. Extracting features from 3D data became main computer vision task. In this paper, we reviewed methods of extracting local features from objects represented by point clouds. The goal of the work was to make theoretical overview and evaluation of selected point cloud detectors and descriptors. We performed an experimental assessment of the repeatability and computational efficiency of individual methods using the well known Stanford 3D Scanning Repository database with the aim of identifying a method which is computationally-efficient in finding good corresponding points between two point clouds. We combine the detectors with several feature descriptors and show which combination of detector and descriptor is suitable for object recognition task in cluttered scenes. Our tests show that choosing the right detector impacts the descriptor's performance in the recognition process. The repeatability tests of the detectors show that the data which contained occlusions have a high impact on their performance. We summarized the results into graphs and described them with respect to the individual tested properties of the methods.

**Keywords:** 3D detector · 3D descriptor · Point cloud · Feature extraction

## 1 Introduction

With easily accessible 3D point cloud acquisition tools it is natural that the field of point cloud processing gained a lot of attention in the past decade. Many computer vision tasks working with 3D point clouds benefit from using 3D local features. Among possible applications of 3D local features are point cloud registration (pose estimation) [6], automatic object localization and recognition in 3D point cloud scenes [26], model retrieval [3] or face recognition [15].

© Springer Nature Switzerland AG 2020
L. J. Chmielewski et al. (Eds.): ICCVG 2020, LNCS 12334, pp. 188–200, 2020.
https://doi.org/10.1007/978-3-030-59006-2_17

The pipeline of these applications usually starts with detecting keypoints using detectors designed by studying the geometric properties of local structures [6]. The next step is the description of the keypoints and finally the classification or matching of the keypoints. The latest deep learning approach combines all three steps into one. The features and their descriptions are learned during the training of the classifier [13, 17].

When using the hand crafted features we need to ensure that the detected keypoints are representative and the descriptors are robust to clutter and occlusion in order to get reliable results (classification, localization,...).

The goal of the paper is to provide the performance evaluation of hand crafted feature detectors for point clouds. First we address the ability to detect repeatable points. Second, since the efficiency of the descriptors can influence the results, we combine the detectors with several feature descriptors and show which combination of detector and descriptor is suitable for object recognition task in cluttered scenes.

## 2   Related Work

Many existing works test and compare methods to extract 3D local features. Filipe and Alexandre [7] work with a real dataset of different RGB-D objects. They test 3D detectors in the Point Cloud Library (PCL) [20] that work directly on RGB-D data. They only evaluate the performance of the detectors in terms of various transformations. Tombari, Salti and di Stefano [25] perform in-depth testing of state-of-the-art detectors, which they divide into two new categories (fixed-scale, adaptive-scale). Testing is performed on 3D scene built by randomly rotating and translating the selected 3D models of synthetic data. A quantitative comparison of the performance of the three most general detectors and two descriptors for monitoring plant health and growth is given in [1]. The choice of test methods was based on their most common use in various applications. Mian, Bennamoun and Owens [16] propose a multi-scale keypoint detection algorithm for extracting scale invariant local features. They also propose a technique for automatically selecting the appropriate scale at each keypoint and extract multi-scale and scale invariant features. They test the methods for 3D object retrieval from complex scenes containing clutter and occlusion. Hänsch, Weberand and Hellwich [10] explore the benefits of 3D detectors and descriptors in the specific context of point cloud fusion that can be used, for example, to reconstruct a surface.

Recent trends in data driven approaches encouraged to employ deep learning. Representative works include PPFNet [5], PCPNet [9], 3D-FeatNet [27] and PPF-FoldNet [4]. All the mentioned works surpassed the handcrafted alternatives with a big margin. 2D features are usually supplemented by useful information about the local orientation, derived from the local visual part of the image. For 3D features, it is very difficult to find a unique and consistent local coordinate frame. None of the aforementioned works were able to attach the local orientation information to 3D patches.

Inspired by [25], we propose a comparison of 3D detectors that work primarily on the point clouds without the use of the color information. Like [25], we focus on the object recognition scenario, which is characterized mainly by a large occlusion. Subsequently, we assess the performance of the selected 3D descriptors in combination with the tested detectors because, it is not clear how different keypoints detectors and descriptors work together. Our purpose is to show the performance of selected combinations in recognizing objects in a scene, as we have not found these combinations in the existing papers.

## 3   3D Keypoint Detectors, Descriptors and Matching

In this section we will describe detectors and descriptors we have used in our tests together with the matching method we have used.

### 3.1   3D Keypoint Detectors

**Harris3D.** The Harris3D detector [22] is the extension of the 2D corner detection method of Harris and Stephens [11]. Harris3D is using first order derivatives along two orthogonal directions on the surface. The derivatives are obtained so that the neighboring area is approximated by quadratic surface. Afterwards, similarly to 2D detector, the covariance matrix is calculated, but the image gradients are replaced by the surface normals. For final detection of keypoints, the Hessian matrix of intensity is used for every point, smoothed by an isotropic Gaussian filter.

**Intrinsic Shape Signatures 3D (ISS3D).** The ISS3D keypoint detector [28] relies on measuring the local distances or the points density. It uses the eigenvalues and eigenvectors of the scatter matrix, created from supporting region points. The eigenvector corresponding to the smallest eigenvalue is used as a surface normal and the ratio between two successive eigenvalues is used to reject similar points.

**Normal Aligned Radial Feature (NARF).** Detector NARF [23] is designed to choose the keypoints, so that they are able to repair the information about the object borders and its surface to obtained highest robustness of the method. The selected points are supposed to be in positions where the surface is stable and where there are sufficient changes in the local neighborhood to be robustly detected in the same place even if observed from different perspectives. At first, the detector determines the object border, for which the change of distances between two adjacent points is used. To achieve the keypoint stability, for every point and its local neighborhood, the amount and dominant direction of the surface changes at this position are determined. The interest value is calculated as the difference between orientations in area and changes of the surface. Afterwards, the smoothing of the interest values and non-maximum suppression is performed to determine the final keypoints.

**SIFT3D.** SIFT 3D is an extension of the original 2D SIFT detector, where for finding the keypoints the scale pyramid of difference of Gaussians (DoG) is used [14]. SIFT 3D is using 3D version of Hessian matrix, where the density function is approximated by sampling the data regularly in space. Over the density function a scale space is built, where search for local maxima of the Hessian determinant is performed. The input cloud is convolved with a number of Gaussian filters whose standard deviations differ by a fixed scale factor. Using the convolution the point clouds are smoothed and the difference is calculated to obtained three to four DoG point clouds. These two steps are repeated to get higher number of DoG in the scale-space. The 3D SIFT keypoints are identified as the scale-space extrema of the DoG function similarly to 2D SIFT.

### 3.2   3D Local Features Descriptors

Many existing 3D local features descriptors use histograms to represent different characteristics of the local surface. Principally, they describe the local surface by combining geometric or topological measurements into histograms according to point coordinates or geometric attributes. We classify these descriptors into two groups: *geometric attribute histogram* and *spatial distribution histogram* based descriptors.

**Geometric Attribute Histogram Based Descriptors.** These descriptors represent local surface by generating histograms based on geometric attributes, as normals or curvatures, of the points on the surface. These descriptors also mostly construct the Local Reference Frame (LRF) and/or Local Reference Axis (LRA).

**Point Feature Histogram (PFH):** The PFH descriptor [19] works with point pairs in the support region. First, the Darboux frame is defined by using the surface normals and point positions, for each pair of the points in the support region. Then, four features for each point pair using the Darboux frame, the surface normals, and their point positions are calculated. The PFH descriptor is generated by accumulating points in particular bins along the four dimensions. The length of the final PFH is based on the number of histogram bins along each dimension.

**Fast Point Feature Histogram (FPFH):** The FPFH descriptor [18] consists of two steps. In the first step, a Simplified Point Feature Histogram (SPFH) is generated for each point by calculating the relationships between the point and its neighbors. In SPFH, the descriptor is generated by chaining three separate histograms along each dimension. In the second step, FPFH descriptor is constructed as the weighted sum of the SPFH of the keypoint and the SPFHs of the points in the support region. The length of the FPFH descriptor depends on the number of histogram bins along each dimension.

**Signature of Histograms of OrienTations (SHOT):** The SHOT descriptor [21] originated as an inspiration by SIFT descriptor. The descriptor encodes the histograms of the surface normals in different spatial locations. First, the LRF is constructed for each keypoint and its neighboring points in the support region are aligned with the LRF. The support region is divided into several units along the radial, azimuth and elevation axes. Local histogram for each unit is generated by accumulating point counts into bins according to the angles between the normals at the neighboring points within the volume and the normal at the keypoint. The final descriptor is chaining all the local histograms.

**Spatial Distribution Histogram Based Descriptors.** These descriptors represent local surface by generating histograms according to spatial coordinates. They mostly start with the construction of a LRF or LRA for keypoint, and divide 3D support region to the LRF/LRA. Afterwards, they generate a histogram for local surface by accumulating the spatial distribution measurements for each spatial bin.

**3D Shape Context (3DSC):** 3DSC descriptor [8] is extension of the 2D Shape Contexts descriptors [2]. The algorithm use the surface normal at the keypoint as its LRA. First, a spherical grid is placed at the keypoint, which is the center of the grid, with the north pole of the grid being aligned with the surface normal. The support region is divided into several bins, logarithmic along the radial dimension and linear along the azimuth and the elevator dimensions of the spherical grid. The 3DSC descriptor is generated by counting the weighted number of points falling into each bin of the 3D grid. The final descriptor is chaining the three partial descriptors, that represent the numbers of bins along the radial, azimuth and elevation axes.

**Unique Shape Context (USC):** USC descriptor [24] is extension of 3DSC, which bypasses computing multiple descriptors at a given keypoint. LRF is constructed for each keypoint and aligned with the local surface in order to provide invariance to rotations and translations. The support region is divided into several bins. Final USC descriptor is generated analogous to the approach used in 3DSC.

### 3.3   Descriptors Matching

In our experiments the objects in the scenes are recognized using the nearest neighbor distance ratio matching method (NNDR). In the nearest neighbor matching method, two descriptors match, if they are the closest and their distance is below a threshold. With this strategy, there is at most one match but it might not be correct, especially when there are few keypoints with similar geometric properties in the point cloud. In this case, the distances from the nearest and the second nearest descriptors are similar. To overcome this property we use the NNDR matching, where the descriptors $D_A$ and $D_B$ are matched if

$$||D_A - D_B||/||D_A - D_C|| < t, \tag{1}$$

where $D_B$ is the first and $D_C$ is the second nearest neighbor to $D_A$ and $t$ is a specified threshold.

# 4    Testing

We tested the performance of detectors and descriptors separately in combination with the tested detectors. In this section we provide the evaluation criteria we have used for the tests and the results we have obtained.

## 4.1    Dataset

For performing our experiments, we selected four models well known in community of 3D point clouds researchers (the Stanford Bunny, the Happy Budha, the Dragon, the Armadillo) from the publicly available 3D model database the Stanford 3D Scanning Repository [12]. All the selected models were scanned by a Cyber-ware 3030 MS scanner. For the testing we used the models individually and scaled them using the scale of 6, 9, 12, 15 times the cloud resolution. We also combined several transformed models to create 15 testing scenes having 265–290 thousands of points. The scenes were created so that different parts of individual models were occluded. The occlusion was from 30% to 60%.

## 4.2    Evaluation Criteria

A keypoint can be useful for identifying an object, when it is detectable under changing conditions including viewpoint change, noise, partial neighborhood occlusion, etc. This property is called repeatability.

A keypoint is repeatable, if it is found in a small neighborhood of the expected location in the scene. Let $k_M$ be a keypoint found on the original model. The expected location in the scene is then $T(k_M)$, where $T()$ is the transformation of the original model (rotation and translation) when inserted into the scene. Let $k_S$ be the keypoint in the scene, which is the closest to $T(k_M)$. The keypoint $k_M$ is repeatable, when

$$\|T(k_M) - k_S\| < \varepsilon. \tag{2}$$

As in [25] we evaluate the *absolute* and *relative repeatability* of the keypoints. When $R$ is the set of repeatable keypoints found on a given model in a given scene, the absolute repeatability is the number of the repeatable keypoints

$$r_a = |R| \tag{3}$$

and the relative repeatability is measured relatively to the number of keypoints found on the model in the scene having a corresponding keypoint on the original model ($K_S$)

$$r_r = \frac{|R|}{|K_S|}. \tag{4}$$

While testing the detectors, the scale was set to 6, 9, 12, 15 times the cloud resolution. The results are averaged over the scales and the scenes.

The *descriptiveness* of selected descriptors is evaluated by combining them with the tested detectors. The purpose is to demonstrate the effect of different detectors on the performance of the feature descriptors. The performance of each

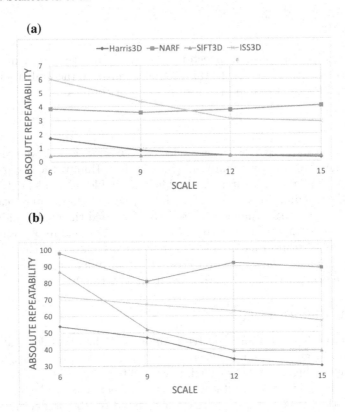

**Fig. 1.** (a) The average absolute repeatability of tested detectors with 60% occlusion in the scene (b) The average absolute repeatability of tested detectors with 30% occlusion in the scene

detector-descriptor combination is also measured by the Precision and Recall generated as follows. First, the keypoints are detected from all the scenes and all the models. A feature descriptor is then computed for each keypoint (not just repeatable points) using the selected descriptors. Next, we use NNDR to perform feature matching of the models with the scenes. If the distance between the pair of keypoint descriptors is less than half of the support radius, which is the threshold, the match is assumed correct. Otherwise, it is a false match. The *precision* is calculated as

$$\text{precision} = \frac{\text{Nr. of correct matches}}{\text{Nr. of identified matches}}. \tag{5}$$

The *recall* is calculated as

$$\text{recall} = \frac{\text{Nr. of correct matches}}{\text{Nr. of corresponding features}}. \tag{6}$$

We also evaluate the *computational effectiveness* of the detectors and descriptors. We calculate the time needed for detecting the keypoints on several objects

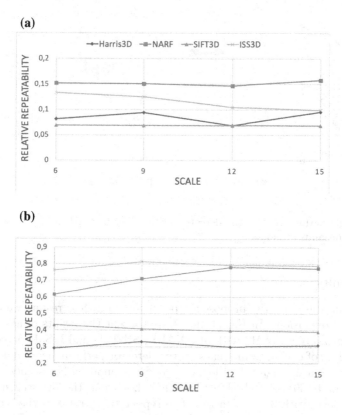

**Fig. 2.** (a) The average relative repeatability of tested detectors with 60% occlusion in the scene (b) The average relative repeatability of tested detectors with 30% occlusion in the scene

or scenes with various point density to determine the effectiveness of detector. Also, we calculate the time needed to describe the extracted keypoints.

All methods are used as implemented in the Point Cloud Library (PCL), which is an open-source library of algorithms for 2D/3D image and point cloud processing. For all the selected detectors, we use the default parameters proposed in the original publications.

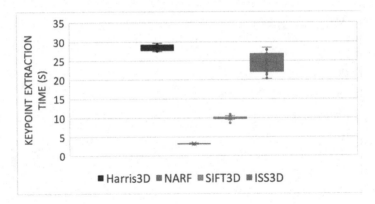

**Fig. 3.** Time performance of the detectors. Scale is kept fixed to 6. The scenes have 265–290 thousands of points

## 4.3  Results

**Detectors.** Based on the proposed experiments and their results we can draw interesting conclusions. In Figs. 1 and 2, we present the results of absolute and relative repeatability of HARRIS3D, NARF, SIFT3D and ISS3D detectors. To test the effect of occlusion in scenes on the detector performance, we investigate scenes that contain highly and low occluded test models. For scenes with high occlusion, up to 60% of each object is occluded, whereas the low occlusion scenes contain objects with 30% occlusion. The repeatability rates in the scenes with high occlusion are very low. Nevertheless, we can see that NARF and ISS3D detectors achieve higher rates than HARRIS3D and SIFT3D in both scenarios. The only exception is the absolute repeatability rate of SIFT3D for scale 6 in low occlusion scenes.

In the highly occluded scenes is the relative repeatability of all detectors very low. An the average number of detected keypoints does not exceed 6.

In the scenes with lower occlusion the relative repeatability reaches 80% for NARF and ISS3D detectors, which means that the points are highly repeatable, while the absolute number of detected points is also high. For HARRIS3D and SIFT3D even though the absolute numbers of detected keypoints is high for scales 6 and 9, the relative repeatability rate reaches only 40%.

The error-bar chart in Fig. 3 reports the measured detection time (in seconds) of methods computed on each model. Because all methods are implemented in the same framework, their performance can be compared by taking the average detection time on the same set of objects. The detection time is almost constant over scenes with varying number of points. The only exception is the ISS3D, where the trend is constant with higher variance.

**Descriptors.** In Fig. 4 we present the precision and recall measures of the descriptor matching task on scale 6 objects. All detector–descriptor combinations achieve more than 65% recall, and 50% precision (except SIFT3D–USC with 38%). We can see that the USC descriptor causes the worst performance regardless of the detector and the SIFT3D detector is the weakest from all when combined with any descriptor. The best performance was achieved by the FPFH descriptor for all the detectors. The performance of PFH, 3DSC and SHOT descriptors varies with the used detectors. The order of decreasing performance is PFH, SHOT and 3DSC descriptors for HARRIS3D, NARF and ISS3D detectors and the other way round for the SIFT3D detector. The best performance overall is achieved by the pair ISS3D–FPFH.

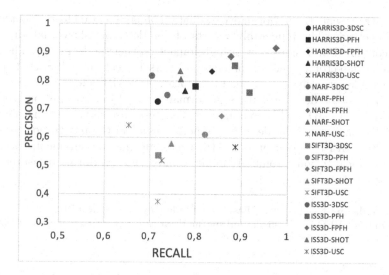

**Fig. 4.** Performance of the selected descriptors combined with different 3D keypoint detectors

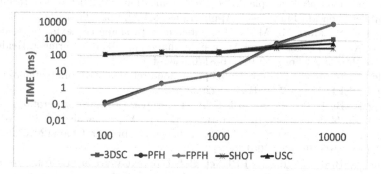

**Fig. 5.** Average time to generate a feature descriptor for a local surface with different number of points in the support region. The scale of both axes is logarithmic

The graph in Fig. 5 reports the measured average time (in milliseconds) of keypoint description according to the number of points in its local region. As with the detectors, the descriptors are implemented in the same framework, so their performance can be compared by taking the average time needed to generate the descriptor of the same model represented in different scales.

We can see that for small-sized support regions (100–1000 points) the PFH a FPFH descriptors are orders of magnitude faster than the other descriptors, but for 10000, they are 10 times slower. All descriptors are comparably fast for neighborhoods with around 5000 points.

## 5    Conclusions

This paper presents the performance evaluation of hand crafted feature detectors for point clouds. The repeatability tests of the detectors show that the data which contained occlusions have a high impact on their performance. The best repeatable detector of the set of test detectors in scale 6 is ISS3D. For the scenes with lower occlusion are the best repeatable NARF and ISS3D detectors. The best detector–descriptor pairs performance for object recognition scenario is achieved by ISS3D–FPFH combination. Our tests also show that choosing the right detector impacts the descriptor's performance in the recognition process. The FPFH ann PFH descriptors are the fastest in time measurements for small number of local region's point. At higher points, the descriptors measurements are comparable.

**Acknowledgment.** This work has been funded by Slovak Ministry of Education under contract VEGA 1/0796/00 and by the Charles University grant SVV-260588.

## References

1. Azimi, S., Lall, B., Gandhi, T.K.: Performance evalution of 3D keypoint detectors and descriptors for plants health classification. In: 2019 16th International Conference on Machine Vision Applications (MVA), pp. 1–6. IEEE (2019)
2. Belongie, S., Mori, G., Malik, J.: Matching with shape contexts. In: Krim, H., Yezzi, A. (eds.) Statistics and Analysis of Shapes, pp. 81–105. Springer, Heidelberg (2006). https://doi.org/10.1007/0-8176-4481-4_4
3. Bold, N., Zhang, C., Akashi, T.: 3D point cloud retrieval with bidirectional feature match. IEEE Access **7**, 164194–164202 (2019)
4. Deng, H., Birdal, T., Ilic, S.: PPF-FoldNet: unsupervised learning of rotation invariant 3D local descriptors. In: Ferrari, V., Hebert, M., Sminchisescu, C., Weiss, Y. (eds.) ECCV 2018. LNCS, vol. 11209, pp. 620–638. Springer, Cham (2018). https://doi.org/10.1007/978-3-030-01228-1_37
5. Deng, H., Birdal, T., Ilic, S.: PPFNet: global context aware local features for robust 3D point matching. In: Proceedings of the IEEE Conference on Computer Vision and Pattern Recognition, pp. 195–205 (2018)
6. Deng, H., Birdal, T., Ilic, S.: 3D local features for direct pairwise registration. In: Computer Vision and Pattern Recognition (CVPR). IEEE (2019)

7. Filipe, S., Alexandre, L.A.: A comparative evaluation of 3D keypoint detectors in a RGB-D object dataset. In: 2014 International Conference on Computer Vision Theory and Applications (VISAPP), vol. 1, pp. 476–483. IEEE (2014)
8. Frome, A., Huber, D., Kolluri, R., Bülow, T., Malik, J.: Recognizing objects in range data using regional point descriptors. In: Pajdla, T., Matas, J. (eds.) ECCV 2004. LNCS, vol. 3023, pp. 224–237. Springer, Heidelberg (2004). https://doi.org/10.1007/978-3-540-24672-5_18
9. Guerrero, P., Kleiman, Y., Ovsjanikov, M., Mitra, N.J.: PCPNet learning local shape properties from raw point clouds. In: Computer Graphics Forum, vol. 37, pp. 75–85. Wiley Online Library (2018)
10. Hänsch, R., Weber, T., Hellwich, O.: Comparison of 3D interest point detectors and descriptors for point cloud fusion. ISPRS Ann. Photogramm. Remote Sens. Spat. Inf. Sci. **2**(3), 57 (2014)
11. Harris, C.G., Stephens, M., et al.: A combined corner and edge detector. In: Proceedings of the 4th Alvey Vision Conference, pp. 23.1–23.6 (1988)
12. Levoy, M., Gerth, J., Curless, B., Pull, K.: The Stanford 3D scanning repository (2005). http://www-graphics.stanford.edu/data/3Dscanrep
13. Li, J., Chen, B.M., Lee, G.H.: SO-Net: Self-organizing network for point cloud analysis. In: 2018 IEEE/CVF Conference on Computer Vision and Pattern Recognition, pp. 9397–9406, June 2018
14. Lowe, D.G.: Object recognition from local scale-invariant features. In: Proceedings of the Seventh IEEE International Conference on Computer Vision, vol. 2, pp. 1150–1157. IEEE (1999)
15. Markuš, N., Pandžić, I., Ahlberg, J.: Learning local descriptors by optimizing the keypoint-correspondence criterion: applications to face matching, learning from unlabeled videos and 3D-shape retrieval. IEEE Trans. Image Process. **28**(1), 279–290 (2019)
16. Mian, A., Bennamoun, M., Owens, R.: On the repeatability and quality of keypoints for local feature-based 3D object retrieval from cluttered scenes. Int. J. Comput. Vis. **89**(2–3), 348–361 (2010). https://doi.org/10.1007/s11263-009-0296-z
17. Qi, C.R., Yi, L., Su, H., Guibas, L.J.: PointNet++: deep hierarchical feature learning on point sets in a metric space. In: Guyon, I., et al. (eds.) Advances in Neural Information Processing Systems, vol. 30, pp. 5099–5108. Curran Associates, Inc. (2017)
18. Rusu, R.B., Blodow, N., Beetz, M.: Fast point feature histograms (FPFH) for 3D registration. In: 2009 IEEE International Conference on Robotics and Automation, pp. 3212–3217. IEEE (2009)
19. Rusu, R.B., Blodow, N., Marton, Z.C., Beetz, M.: Aligning point cloud views using persistent feature histograms. In: 2008 IEEE/RSJ International Conference on Intelligent Robots and Systems,. pp. 3384–3391. IEEE (2008)
20. Rusu, R.B., Cousins, S.: 3D is here: Point Cloud Library (PCL). In: IEEE International Conference on Robotics and Automation (ICRA), Shanghai, China, 9–13 May 2011
21. Salti, S., Tombari, F., Di Stefano, L.: SHOT: unique signatures of histograms for surface and texture description. Comput. Vis. Image Underst. **125**, 251–264 (2014)
22. Sipiran, I., Bustos, B.: Harris 3D: a robust extension of the Harris operator for interest point detection on 3D meshes. Vis. Comput. **27**(11), 963 (2011). https://doi.org/10.1007/s00371-011-0610-y
23. Steder, B., Rusu, R.B., Konolige, K., Burgard, W.: Point feature extraction on 3D range scans taking into account object boundaries. In: 2011 IEEE International Conference on Robotics and Automation (ICRA), pp. 2601–2608. IEEE (2011)

24. Tombari, F., Salti, S., Di Stefano, L.: Unique shape context for 3D data description. In: Proceedings of the ACM Workshop on 3D Object Retrieval, pp. 57–62 (2010)

25. Tombari, F., Salti, S., di Stefano, L.: Performance evaluation of 3D keypoint detectors. Int. J. Comput. Vis. **102**(1–3), 198–220 (2013). https://doi.org/10.1007/s11263-012-0545-4

26. Vargas, J., Garcia, A., Oprea, S., Escolano, S., Rodriguez, J.: Object recognition pipeline: Grasping in domestic environments, pp. 18–33. IGI Global (2018)

27. Yew, Z.J., Lee, G.H.: 3DFeat-Net: weakly supervised local 3D features for point cloud registration. In: Ferrari, V., Hebert, M., Sminchisescu, C., Weiss, Y. (eds.) ECCV 2018. LNCS, vol. 11219, pp. 630–646. Springer, Cham (2018). https://doi.org/10.1007/978-3-030-01267-0_37

28. Zhong, Y.: Intrinsic shape signatures: a shape descriptor for 3D object recognition. In: 2009 IEEE 12th International Conference on Computer Vision Workshops (ICCV Workshops), pp. 689–696. IEEE (2009)

# A New Watermarking Method for Video Authentication with Tamper Localization

Yuliya Vybornova$^{(\boxtimes)}$ (iD)

Samara National Research University, Samara, Russia
vybornovamail@gmail.com

**Abstract.** In this paper, a new method for video authentication is proposed. The method is based on construction of watermark images, which serve as a secondary carrier for the binary sequence. A unique watermark image is embedded into the coefficients of Discrete Wavelet Transform of each video frame. The analysis of images extracted from video allows to detect spatial attacks, and the sequence carried by the extracted images provides the ability to determine the type of temporal attack and localize the frames, which are tampered. The experimental study on the method quality and efficiency is conducted. According to the results of experiments, the method is suitable for solving authentication tasks. Furthermore, the method is robust to compression and format re-encoding.

**Keywords:** Digital watermarking · Authentication · Video protection · Tamper localization · Forgery detection · Forensics · Discrete Wavelet Transform · MPEG-4 · Motion JPEG

## 1 Introduction

Today, the number of tasks requiring video content authentication is growing steadily. The ability to prove the video authenticity is especially important when it is used as evidence of human actions, for example, in court. In most cases, it is difficult to guarantee that the digital video obtained as evidence is exactly the one that was actually captured by the camera.

The video authentication system ensures the integrity of digital video and checks, whether the video is tampered or not. But in most cases, such systems do not provide the information about the type of attack, and video is just considered inapplicable as evidence. A video can be modified by replication, deletion, insertion and replacement of frames or objects in frames. Thus, the two common types of video tampering attacks are spatial (i.e. modification of frame content) and temporal (i.e. modification of frame order).

The authentication of digital video can be performed both by means of cryptography, such as digital signatures and hash functions, and by means of watermarking. Such methods belong to the class of active protection, that is, protective measures are performed before attempting an attack.

© Springer Nature Switzerland AG 2020
L. J. Chmielewski et al. (Eds.): ICCVG 2020, LNCS 12334, pp. 201–213, 2020.
https://doi.org/10.1007/978-3-030-59006-2_18

In most cases, the cryptographic primitives only provide data integrity verification without the ability to localize the tampered fragment. For example, in [1] hash is calculated using Discrete Cosine Transform and Singular Value Decomposition. In [2] a blockchain model is proposed, which uses both hashes and signatures to verify the video integrity. However, some methods like proposed in [3] do provide the ability of tamper localization. The authors propose to form hashes robust to temporal attacks but allowing to localize the tampered content of the video frame.

Another direction of forgery detection is represented by methods of passive protection, which can determine the fact of malicious modifications by analyzing the statistical properties of the data. These methods are typically implemented via machine learning tools. Generally, the efficiency of such methods depends on the particular task, video content, and codecs used for compression. In addition, most existing methods are very sensitive to camera settings and lighting conditions [4].

Thus, the task of this research is to create a universal method that allows to determine and localize both temporary and spatial attacks, and at the same time does not depend on video content and format. Semi-fragile digital watermarking seems to be the best solution since it allows to avoid disadvantages mentioned above for other techniques.

The rest of the paper is organized as follows. Section 2 overviews the related work. In Sect. 3, the proposed approach for video watermarking is presented. Section 4 comprises experimental study on quality and efficiency of the proposed method. Section 5 provides general conclusions and the main issues of the future work.

## 2  Related Work

The main idea of most existing video watermarking methods is to embed the protective information into frames by introducing distortions that are acceptable in terms of accuracy and usually invisible to the legitimate user.

Generally, the watermark is defined by a bit vector of a given length or a bitmap image. In the first case, it is hard to ensure the enough robustness providing the integrity of the embedded bits. In the second case, the embedded image can also be distorted even when legitimate modifications. For this reason, the process of watermark verification requires the calculation of bit error rate (BER) or normal correlation (NC). Such approach to verification is suitable in the task of copyright protection, but when verifying video authenticity, it can be hard to realize whether the concrete values of BER/NC are acceptable, or they should be considered as a confirmation of attack.

The most common classification divides existing watermarking methods into two groups depending on the embedding domain. In the spatial domain, the watermark is embedded into the video frame by directly changing the brightness values of pixels, while in the transform domain, the watermark is introduced by changing the decomposition coefficients of the image matrix. Most research in the field of video watermarking focuses on the second group of methods, i.e. methods of embedding in the transform domain, since they demonstrate high robustness, and at the same time introduce less visible distortion than methods of embedding in the spatial domain. The cost for these advantages is a high computational complexity in comparison with watermarking in the spatial domain.

The discrete wavelet transform (DWT), discrete cosine transform (DCT), and singular value decomposition (SVD) are the most commonly used today. DCT is more often used in steganography and less often in watermarking. In [5] DCT is applied in combination with DWT and SVD, and in [6] authors use discrete sine transform instead.

Applying SVD in combination with various transforms is a very spread technique, since it provides a very high robustness of the embedded information. Thus, in [7] SVD is applied to one of DWT sub-bands; and in [8] SVD is applied to the watermark image, which is then embedded into DWT sub-band. In [9] authors propose to use SVD as a domain for the encrypted contourlet coefficients. The main drawback of most methods using SVD is that they require the original video, i.e. they are non-blind. However, there exist some methods allowing to perform extraction from SVD without using the host signal, e.g. [10], but the informational capacity of such methods is much lower compared to non-blind ones.

In video and image watermarking, the most common transform domain is the space of DWT coefficients, since it allows to preserve quality of the watermarked data, as well as provides high watermark capacity and robustness [11].

The most used and the simplest wavelet decomposition is Haar wavelet transform. According to the research provided in [12], the Haar DWT provides the highest quality of the extracted information compared to other wavelet decompositions. As already mentioned above, DWT can be combined with DCT and/or SVD for increasing the watermarking quality [5,7,8]. However, there are a variety of another techniques using DWT. Thus, for example, in [13] authors propose to embed different fragments of a watermark into different scenes of the video, while for motionless frames the same watermark fragments serve as embedding information. In [14] authors proposed to increase robustness by embedding short messages only into the region of interest calculated on the basis of features extracted from the carrier video. The watermark is embedded into various bit planes of 2-level DWT sub-bands.

Furthermore, video watermarking methods can be classified into format-oriented and robust to transcoding. The methods of the first type can be applied only to particular video formats. In recent years, a variety of methods were proposed for MPEG-4 standard. Despite the fact that MP4 format is one of the most known nowadays, the problem of its protection is still actual. In [15] the watermarking process is based on the Chinese Remainder Theorem. The watermark images are hard to verify because of the distortions introduced by the embedding procedure. Thus, the method is not suitable for authentication purposes. In [16] authors proposed a robust to recompression watermarking scheme based on chromatic DCT. The scheme utilizes major features of H.246/AVC coding standard and allows to detect frame tampering. In [17] authentication information is constructed not only on the basis of the video features, but also using the audio content of MP4 file, and then embedded in subtitles. This scheme allows to detect frame addition and removal. In [18] the video hash is encrypted and embedded into audio data, and the encrypted audio hash is embedded into synchronization information of MP4 file. The proposed scheme is robust against compression and can be applied for tamper detection.

Transcoding invariant methods are mostly refer to robust watermarking. For example, the mentioned above methods [8,9], which use SVD, allow to transcode the video into the other formats. In [19] a watermark image of a small size is embedded into a given area of a key frame. In this approach, authentication requires manual selection of parameters, until the watermark becomes clearly visible.

As for authentication tasks, [17] allows to detect temporal attacks, but can be applied only for video with subtitles. In [20] the authors determine the specific frames among the non-motion frames, and then apply 3D-DWT for embedding. The approach is not resistant to temporal attacks, which means it can be used only to detect them, not to localize. Some methods, like [16] provide the detection of spatial attacks, as well as tamper region localization. In [21] authors propose a fragile watermarking scheme using Least Significant Bit (LSB) embedding strategy, which allows to detect and localize changes on the frame.

Thus, the development of a video authentication method for detection and localization of both spatial and temporal attacks, and applicable for common video formats, is still an actual task, which is solved in this paper.

## 3 Proposed Method

### 3.1 Watermark Generation

In this paper, two types of watermarks are considered. The first is a watermark sequence represented by a set of unique binary subsequences $s_i$ of a fixed length $l$. The length $L$ of the whole sequence $S = s_1, s_2, \ldots, s_K$ depends on the number of frames $K$ in the carrier video, and can be calculated as $L = l \times (K \bmod (2^l - 1))$.

The watermark sequence can be produced by a random permutation $(\sigma(b_1), \sigma(b_2), \ldots)$ of all possible templates $b_j$ of length $l$ and taking the first $K$ elements $(\sigma(b_1), \sigma(b_2), \ldots, \sigma(b_K))$. Here, $j = \overline{1, |\mathbf{B}_l|}$ and $|\mathbf{B}_l| = 2^l - 1$ is a size of a set of all $b_j$ of length $l$ (note, that zero-sequence is excluded).

The second watermark type is a noise-like image used as a secondary carrier for a subsequence $s_i$. The process of noise-like image construction is shown in detail in one of our previous works [22].

To obtain a noise-like image, its two-dimensional spatial spectrum is formed by arranging the bits of the corresponding watermark sequence in the spectral domain. With an equal angle step, two-dimensional impulses are placed on two rings of different radii $r$ and $r + \Delta r$, i.e. depending on the value of the corresponding bit of the watermark sequence, the impulse is placed on the ring of a smaller or larger radius. Then, using the inverse discrete Fourier transform (DFT), a transition to a two-dimensional image is performed.

However, in this study the above described method is slightly modified by adding some new distinguishing features to improve robustness and security of the sequence carried inside the noise-like image.

To avoid the watermark image tampering by reproducing the sequence of noise-like images there is a possibility to construct different spectrums and, consequently, different watermark images for the same $s_i$ by setting the values of impulses at random both for real and imaginary parts of the spectrum. Such a modification allows to prevent the attempt of the adversary to construct the same sequence of watermark images and embed it into the tampered video.

Figure 1a and Fig. 1b demonstrate the example of two different watermarks constructed for the same sequence $s = (1, 0, 0, 1, 1, 1, 0, 0, 1, 1)$ with parameters $N = 512, r = 10, \Delta r = 8$.

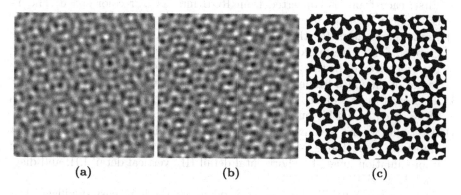

(a)                              (b)                              (c)

**Fig. 1.** Noise-like images: (a), (b) halftone; (c) after binarization

The one more change concerns the number of bits per pixel. The method proposed in this paper operates better with binarized noise-like images. This is explained by the fact that video encoding (especially MPEG-4) introduces considerable distortions into the watermark image. So, the binary image is used since it can be restored much more easily than the halftone one. The example of a noise-like image (shown in Fig. 1a) after binarization is shown in Fig. 1c.

## 3.2 Watermark Embedding

After the construction of the watermark image sequence, each video frame is watermarked by embedding the noise-like image into the wavelet transform domain using LSB-strategy. Such semi-fragile watermarking technique combined with the use of highly robust noise-like watermarks provides robustness against video format change including re-encoding with low compression ratio.

The video $V$ is processed frame-by-frame, and it should be noted that for each frame the watermark image is unique. If the watermark sequence length

$L < K$, then the video should be divided into $t$ blocks of $K_t \leq L$ frames and each frame should be individually watermarked with sequence $S$.

The use of a unique subsequence $s_i$ for each frame $vf_i$ allows to construct a one-to-one mapping $s_i \leftrightarrow vf_i$ which makes the authentication procedure as easy as possible. By analyzing the extracted sequence and comparing it with the original, the following conclusions regarding the sequence integrity can be made: 1) the absence of the sequence fragment indicates that corresponding frames are dropped; 2) the altered order of $s_i$ is an evidence of frame swapping; 3) repeated $s_i$ is a consequence of frame duplication; 4) zero $s_i$ means that the annular spectrum can not be detected.

Although the binary sequence can be repeated, the sequence of watermark images is unique for each block since the spectrum impulses are randomized at the stage of watermark generation.

The algorithm for watermark embedding is as follows.

1) First, each frame is converted from RGB into YCbCr color space. The Y-component is chosen for the further processing.
2) Next, the location for the watermark is selected, and the Y-component is cropped. The watermark location means an arbitrary $2N \times 2N$ fragment of the video frame. In this paper, the watermark is simply embedded into the center of the video frame (namely, its Y-component).
3) Then, watermark pixels are converted to the range $[0; 15]$, i.e. each pixel is represented by a 4-bit number.
4) After this, frame is converted into wavelet domain using the Haar wavelet transform. The frame is decomposed into four sub-bands of $N \times N$ size: approximation image LL, horizontal detail HL, vertical detail LH, and diagonal detail HH.
5) The watermark pixel values are embedded into the four least significant bits of the chosen sub-band.
6) Next, the inverse DWT is performed.
7) Finally, the initial size of the video frame is restored by adding the cropped fragments of Y-component and combining it with the Cb and Cr chroma components.
8) After transition back to RGB color space, the frames are successively added into resulting video.

### 3.3   Watermark Extraction

The extraction of watermark image sequence consists in performing of the following steps for each frame.

1) RGB videoframe is converted to YCbCr. The $2N \times 2N$ fragment containing the watermark is selected.
2) The DWT is calculated, and the sub-band used when embedding is selected.
3) The four least significant bits of the sub-band are the 4-bit watermark pixel values.

4) The extracted watermark is converted to the range of [0; 255].
5) The median filter is applied to the watermark image to obtain better results for the further extraction of binary watermark sequence.

To extract the binary sequence carried inside each watermark image, it is necessary to calculate the DFT of the image, to localize the coordinates of the spectral components having a large amplitude (impulses), and to estimate radii of the rings.

To find large spectral impulses, a local window of size $3 \times 3$ is used. First, local maxima of DFT module are calculated, and other pixels inside the window are set to zero. Next, the largest $2 \times (l + 2)$ elements are selected from those which remain non-zero.

## 4   Experimental Study

To evaluate the effectiveness of the proposed method, several computational experiments were conducted. A video dataset contains eight uncompressed videos from MCL-V Database [23]. For each video, the size of frame is $1920 \times 1080$ pixels. Number of frames varies from 120 to 180.

The watermark image dataset contains 200 random binarized noise-like images constructed on the basis of a 2000-bit sequence with parameters $l = 10$, $N = 512$, $r = 10$, $\Delta r = 8$.

### 4.1   Quality and Efficiency

This section provides the experimental study regarding the visual indistinguishability of the embedded watermark and the operability of the proposed method. The experiment is aimed at finding balance between the method efficiency and quality of the watermarked video.

The sequence of watermark images was successively embedded into various DWT sub-bands of each video. The output video was encoded into three most common video formats: .mp4 (MPEG-4), .avi (Motion JPEG AVI), .mj2 (Motion JPEG 2000).

As a measure of the method efficiency, the watermark sequence integrity, is considered. It can be estimated as a rate of its accurate extraction, i.e. a ratio of correctly extracted watermark bits to the total length of the watermark sequence. The rate of correct extraction is denoted by $R$.

As a measure of fidelity, the common characteristics of the image quality, peak signal-to-noise ratio (PSNR), is calculated.

The experiment has shown that, when embedding into LL sub-band, the watermark becomes visible. HH sub-band demonstrated high imperceptibility, but its robustness to MPEG-4 encoding is a bit lower compared with other detail sub-bands. The best balance between fidelity and efficiency is obtained for HL and LH sub-bands: the results are presented in Table 1.

**Table 1.** Results of quality assessment

| V | HL sub-band | | | | | | LH sub-band | | | | | |
|---|---|---|---|---|---|---|---|---|---|---|---|---|
| | .mj2 | | .avi | | .mp4 | | .mj2 | | .avi | | .mp4 | |
| | PSNR | R | PSNR | R | PSNR | R | PSNR | R | PSNR | R | PSNR | R |
| BQ | 37.48 | 1 | 37.40 | 1 | 37.09 | 0.98 | 37.40 | 1 | 37.31 | 1 | 37.04 | 0.97 |
| TN | 36.83 | 1 | 36.73 | 1 | 34.46 | 1 | 36.98 | 1 | 36.88 | 1 | 34.54 | 1 |
| SK | 37.41 | 1 | 37.34 | 1 | 35.30 | 0.99 | 37.42 | 1 | 37.35 | 1 | 35.30 | 0.99 |
| OT | 37.29 | 1 | 37.19 | 1 | 37.08 | 0.97 | 37.36 | 1 | 37.26 | 1 | 37.14 | 0.97 |
| KM | 37.08 | 1 | 36.99 | 1 | 36.81 | 1 | 37.11 | 1 | 37.02 | 1 | 36.86 | 1 |
| EA | 37.19 | 1 | 37.11 | 1 | 36.48 | 1 | 37.15 | 1 | 37.06 | 1 | 36.48 | 0.99 |
| CR | 37.53 | 1 | 37.44 | 1 | 36.01 | 0.98 | 37.56 | 1 | 37.46 | 1 | 36.02 | 0.99 |
| BC | 37.02 | 1 | 36.93 | 1 | 35.48 | 0.99 | 37.04 | 1 | 36.95 | 1 | 35.49 | 0.99 |

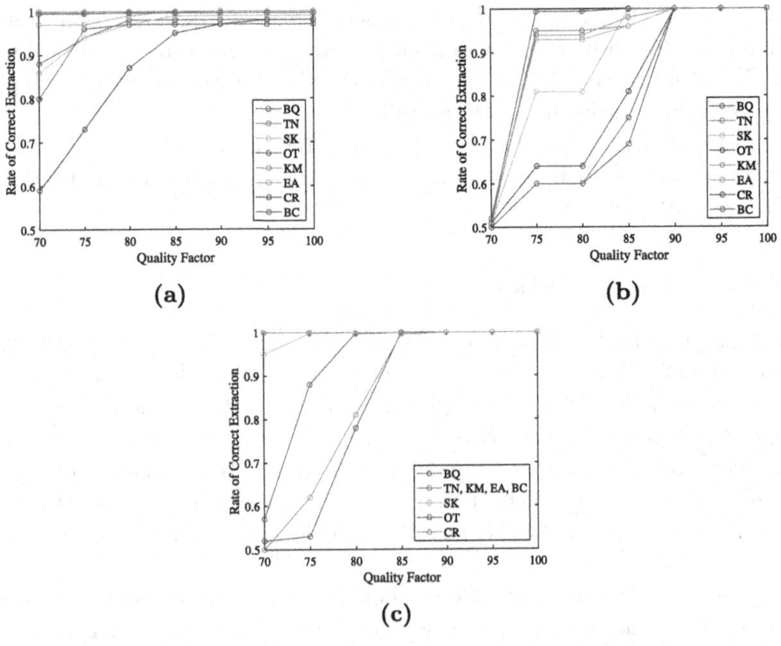

**Fig. 2.** Rate of correct extraction after HL-embedding and encoding with compression: a) MPEG-4; b) Motion JPEG AVI; c) Motion JPEG 2000

## 4.2 Robustness Against Compression

When evaluating the method robustness, the acceptable (legitimate) modifications should be determined. Since the purpose of this paper is to solve the problem of video authentication, the watermark information should remain unchanged when the video is encoded, as well as re-encoded or compressed.

To show the method invariance to encoding with compression, the following experiment was conducted. The watermark was embedded into video, and then the video was encoded with different quality factor $QF$. This procedure was performed for three video formats: .mp4, .avi, .mj2.

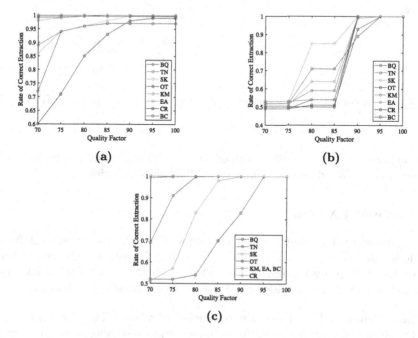

(a)

(b)

(c)

**Fig. 3.** Rate of correct extraction after LH-embedding and encoding with compression: a) MPEG-4; b) Motion JPEG AVI; c) Motion JPEG 2000

The results of the experimental study for various formats in the case of embedding into HL and LH sub-bands are shown in Fig. 2 and Fig. 3 respectively.

According to the results obtained, the watermark carried in the LH-sub-band is less robust to compression, especially for the case of Motion JPEG AVI encoding. For this reason, the next subsections comprise the results only for embedding into HL sub-band.

## 4.3  Robustness Against Re-Encoding

It should be noted that when performing an attack, an adversary can re-encode the video with another quality factor. To analyze the possible effect of this manipulation, each video was firstly watermarked and encoded to three different formats, and then re-encoded to the same format with various values of the quality factor.

The results are shown in Fig. 4. It should be noted that for the case of .mj2 format the results coincide with those from Subsect. 4.2. In this regard, the graphics is not provided.

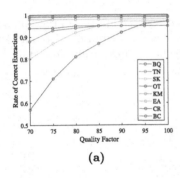
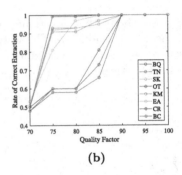

**(a)**                                      **(b)**

**Fig. 4.** Rate of correct extraction after re-encoding with compression: a) MPEG-4; b) Motion JPEG AVI

Compared with the results obtained for one-time encoding in previous subsection, re-encoding does not significantly affect the extraction rate.

### 4.4  Temporal Attacks

The experiment comprises three types of temporal attacks: frame removal, frame reordering and frame addition. Before simulating the attacks, the videos from the dataset were watermarked and encoded to MPEG-4 with $QF = 100$. The attacked video was also encoded to MPEG-4 with highest $QF$ value.

1) Frame Removal. The attack was conducted as follows. The video fragments of a given size were removed from the watermarked video. The first frame of the fragment was selected randomly. The size of the removed fragment varied from 10 to 50% of the total number of video frames. To calculate the extraction rate $R$ between extracted watermark sequence and original, the latter was reconstructed by removal of $s_i$ corresponding to deleted frames.
2) Frame Reordering. To conduct this attack, two random video fragments of a given size were swapped. The size of the fragments varied from 10 to 40% of the total size of the video. To calculate the extraction rate $R$ between extracted watermark sequence and original, the latter was reconstructed by changing the order of $s_i$ in correspondence with swapped video fragments.
3) Frame Addition. For this attack, a random video fragment of a given size was copied and inserted into an arbitrary place of the video. The size of the fragments varied from 10 to 50% of the total size of the video. To calculate the extraction rate $R$, the extracted sequence was reconstructed by removing the sequence fragment extracted from the frames, which were added.

The experiments have shown that authentication quality does not significantly depend on the fraction of changed frames. For this reason, Table 2 provides the results averaged over all cases. According to the results obtained, the watermark sequence is preserved after all types of temporal attacks.

**Table 2.** Extraction rate after temporal attacks

| $V$ | BQ | TN | SK | OT | KM | EA | CR | BC |
|------|------|------|------|------|------|------|------|------|
| Deletion | 0.95 | 0.99 | 0.97 | 0.96 | 0.98 | 0.98 | 0.97 | 0.99 |
| Reordering | 0.96 | 0.99 | 0.97 | 0.96 | 0.99 | 0.99 | 0.96 | 0.99 |
| Addition | 0.96 | 0.99 | 0.98 | 0.96 | 0.98 | 0.98 | 0.97 | 0.99 |

### 4.5   Spatial Attacks

Besides temporal attacks, the proposed method also allows to detect and localize the spatial attacks, i.e. forgery of video frame content. But, here, instead of the binary sequence, the watermark image itself is used to detect forgery.

To demonstrate the idea, the following experiment was conducted. An arbitrary fragment of the video frame was modified by assigning the pixels with random values. After this, the watermark was extracted from the frame. The "tampered" frame is shown in Fig. 5a, while Fig. 5b shows the watermark extracted from the modified frame.

**Fig. 5.** a) Tampered frame of CR video; b) watermark extracted from the tampered frame

It can be clearly seen from the Fig. 5, that the distorted fragment of the watermark coincides with the tampered region of the frame.

However, it should be noted that, forgery detection is possible only for the part of the frame, which carries a watermark. In this study, a watermark was embedded into the central part of a video frame, but generally, any fragment (or several fragments) can be chosen depending on the video content.

## 5   Conclusions and Future Work

In this paper, a new method for video authentication is proposed. The method is based on construction of watermark images, which serve as a secondary carrier for the binary sequence. A unique watermark image is embedded into the coefficients of Discrete Wavelet Transform of each video frame.

The analysis of images extracted from video allows to detect spatial attacks, and the sequence carried by the extracted images provides the ability to determine the type of temporal attack and localize the frames, which are tampered.

The experimental study on the method quality and efficiency is conducted. According to the results of experiments, the method is suitable for solving authentication tasks. Furthermore, the method is robust to compression and format re-encoding.

Future work is supposed to be directed towards the following issues:

1. a development of an algorithm for differentiation of attack types and automatic localization of tampered fragments;
2. an enhancement of the watermark robustness against compression by developing an improved detector for amplitude peaks;
3. a study on robustness against various types of possible geometrical attacks, like cropping and rotation.

**Acknowledgements.** The reported study was funded by RFBR (Russian Foundation for Basic Research): projects No. 19-29-09045, No. 19-07-00474, No. 19-07-00138, No. 20-37-70053.

# References

1. Dabhade, V., Bhople, Y.J., Chandrasekaran, K., Bhattacharya, S.: Video tamper detection techniques based on DCT-SVD and multi-level SVD. In: TENCON IEEE, pp. 1–6 (2015)
2. Ghimire, S., Choi, J., Lee, B.: Using blockchain for improved video integrity verification. IEEE Trans. Multimed. **22**(1), 108–121 (2019)
3. Khelifi, F., Bouridane, A.: Perceptual video hashing for content identification and authentication. IEEE Trans. Circuits Syst. Video Technol. **29**(1), 50–67 (2019)
4. Sitara, K., Babu, M.: Digital video tampering detection: an overview of passive techniques. Digit. Invest. **18**, 8–22 (2016)
5. Aditya, B., Avaneesh, U., Adithya, K., Murthy, A., Sandeep, R., Kavyashree, B.: Invisible semi-fragile watermarking and steganography of digital videos for content authentication and data hiding. Int. J. Image Graph. **19**(3), 1–19 (2019)
6. Shiddik, L., Novamizanti, L., Ramatryana, I., Hanifan, H.: Compressive sampling for robust video watermarking based on BCH code in SWT-SVD domain. In: International Conference on Sustainable Engineering and Creative Computing (ICSECC), pp. 223–227 (2019)
7. Sharma, C., Bagga, A.: Video watermarking scheme based on DWT, SVD, Rail fence for quality loss of data. In: 4th International Conference on Computing Sciences (ICCS), pp. 84–87 (2018)
8. Alenizi, F., Kurdahi, F., Eltawil, A.M., Al-Asmari, A.K.: Hybrid pyramid-DWT-SVD dual data hiding technique for videos ownership protection. Multimed. Tools Appl. **78**(11), 14511–14547 (2018). https://doi.org/10.1007/s11042-018-6723-9

9. Barani, M.J., Ayubi, P., Valandar, M.Y., Irani, B.Y.: A blind video watermarking algorithm robust to lossy video compression attacks based on generalized Newton complex map and contourlet transform. Multimed. Tools Appl. **79**(3), 2127–2159 (2020)
10. Guangxi, C., Ze, C., Daoshun, W., Shundong, L., Yong, H., Baoying, Z.: Combined DTCWT-SVD-based video watermarking algorithm using finite state machine. In: Eleventh International Conference on Advanced Computational Intelligence, pp. 179–183 (2019)
11. Rakhmawati, L., Wirawan, W., Suwadi, S.: A recent survey of self-embedding fragile watermarking scheme for image authentication with recovery capability. EURASIP J. Image Video Process. **2019**(1), 1–22 (2019). https://doi.org/10.1186/s13640-019-0462-3
12. Solanki, N., Khandelwal, S., Gaur, S., Gautam, D.: A comparative analysis of wavelet families for invisible image embedding. In: Rathore, V.S., Worring, M., Mishra, D.K., Joshi, A., Maheshwari, S. (eds.) Emerging Trends in Expert Applications and Security. AISC, vol. 841, pp. 219–227. Springer, Singapore (2019). https://doi.org/10.1007/978-981-13-2285-3_27
13. Sujatha, C.N., Sathyanarayana, P.: DWT-based blind video watermarking using image scrambling technique. In: Satapathy, S.C., Joshi, A. (eds.) Information and Communication Technology for Intelligent Systems. SIST, vol. 106, pp. 621–628. Springer, Singapore (2019). https://doi.org/10.1007/978-981-13-1742-2_62
14. Wagdarikar, A., Senapati, R.: Optimization based interesting region identification for video watermarking. J. Inf. Secur. Appl. **49**, 1–17 (2019)
15. Wagdarikar, A.M.U., Senapati, R.K., Ekkeli, S.: A secure video watermarking approach using CRT theorem in DCT domain. In: Panda, G., Satapathy, S.C., Biswal, B., Bansal, R. (eds.) Microelectronics, Electromagnetics and Telecommunications. LNEE, vol. 521, pp. 597–606. Springer, Singapore (2019). https://doi.org/10.1007/978-981-13-1906-8_61
16. Tian, L., Dai, H., Li, C.: A semi-fragile video watermarking algorithm based on chromatic residual DCT. Multimed. Tools Appl. **79**(5), 1759–1779 (2020)
17. Wong, K., Chan, C., Maung, M.A.: Lightweight authentication for MP4 format container using subtitle track. IEICE Trans. Inf. Syst. **E103.D**(1), 2–10 (2020)
18. Maung, M.A.P., Tew, Y., Wong, K.: Authentication of Mp4 file By perceptual hash and data hiding. Malaysian J. Comput. Sci. **32**(4), 304–314 (2019)
19. Vega-Hernandez, P., Cedillo-Hernandez, M., Nakano, M., Cedillo-Hernandez, A., Perez-Meana, H.: Ownership identification of digital video via unseen-visible watermarking. In: 7th International Workshop on Biometrics and Forensics (IWBF), pp. 1–6 (2019)
20. Cao, Z., Wang, L.: A secure video watermarking technique based on hyperchaotic Lorentz system. Multimed. Tools Appl. **78**(18), 26089–26109 (2019). https://doi.org/10.1007/s11042-019-07809-5
21. Munir, R., Harlili: A secure fragile video watermarking algorithm for content authentication based on arnold cat map. In: 4th International Conference on Information Technology, pp. 32–37 (2019)
22. Vybornova, Y., Sergeev, V.: Method for vector map protection based on using of a watermark image as a secondary carrier. In: ICETE 2019 - Proceedings of the 16th International Joint Conference on e-Business and Telecommunications, pp. 284–293 (2019)
23. Lin, J.Y., Song, R., Wu, C.-H., Liu, T.-J., Wang, H., Kuo, C.-C.J.: MCL-V: a streaming video quality assessment database. J. Vis. Commun. Image Represent. **30**, 1–9 (2015)

# RGB-D and Lidar Calibration Supported by GPU

Artur Wilkowski[1]([✉])[iD] and Dariusz Mańkowski[2]

[1] Institute of Control and Computation Engineering,
Warsaw University of Technology, ul. Nowowiejska 15/19,
00-665 Warsaw, Poland
`artur.wilkowski@pw.edu.pl`
[2] United Robots Sp. z o.o., ul. Świeradowska 47, 02-662 Warsaw, Poland
`mankowski@unitedrobots.co`

**Abstract.** There can be observed a constant increase in the number of sensors used in computer vision systems. This affects especially mobile robots designed to operate in crowded environment. Such robots are commonly equipped with a wide range of depths sensors like Lidars and RGB-D cameras. The sensors must be properly calibrated and their reference frames aligned. This paper presents a calibration procedure for Lidars and RGB-D sensors. A simple inflated ball is used as a calibration pattern. The method applies RanSaC algorithm for pattern detection. The detected sphere centroids are then aligned to estimate rigid transformation between sensors. In addition, an improved GPU RanSaC procedure together with color filtering for RGB-D sensors is used for increased efficiency. The experiments show that this basic calibration setup offers accuracies comparable to those reported in literature. There is also demonstrated a significant speedup due to utilization of GPU-supported procedures as well as proposed color prefiltering.

**Keywords:** Calibration · Lidar · RGB-D sensors · RanSaC · GPGPU computing

## 1 Introduction

The current trend is to equip vision systems with a large number of visual sensors. A leading role in this trend is held by intelligent houses, autonomous cars or mobile robots. This is partially due to application-specific requirements, partially due to increasing affordability of equipment (e.g. introduction of cheap RGB-D) devices, advances in machine vision algorithms or increasing capacity of computer networks.

One area where visual sensors are intensively introduced is mobile robotics. Traditionally, visual sensors are used here to solve SLAM tasks. However, very

The research was supported by the United Robots company under a NCBiR grant POIR.01.01.01-00-0206/17.

L. J. Chmielewski et al. (Eds.): ICCVG 2020, LNCS 12334, pp. 214–226, 2020.
https://doi.org/10.1007/978-3-030-59006-2_19

often new robots are designed to operate in cluttered, uncontrolled environments, where interaction between humans and robots is a normal mode of operation. In such cases additional sensors must be utilized to ensure safety, dynamic obstacle avoidance and efficient human-robot interaction (Fig. 1).

**Fig. 1.** A robot equipped with multiple depth sensors. Top frame: Velodyne VLP-16. Bottom frame: Intel RealSense D435i.

**Fig. 2.** Calibration pattern

Since robots heavily rely on spatial (3D) visual data, it is a common configuration to supply robots with 3D sensors like Lidars or RGB-D cameras, which are complementary technologies. Lidars typically offer long-range and high-precision but sparse scans. Equipment price is considerable here. Much more affordable RGB-D sensors or stereo cameras tend to work well in closer ranges offering much denser coverage of space and color information. In result, robots typically carry both types of sensors (potentially multiplicated). In order to use sensors effectively they must be calibrated. The relation between reference frames of sensors must be established in order to consistently overlay spatial data from different sensors.

In this paper there is proposed a calibration procedure for 3D Lidar and RGB-D sensors. The procedure was designed to take into account the following practical assumptions: 1) the method should be operable in out-of-lab working environment, should use affordable equipment and be relatively easy to carry out, 2) the method should make advantage of hardware features.

Regarding point 1), the proposed calibration procedure uses a single calibration pattern, which is an unicolor inflatable ball. The operator is asked to simply walk in front of a robot carrying the ball (Fig. 2), or conversely the robot can perform autonomous movements with respect to the ball. The detection algorithms are designed to work on-line in order to provide immediate feedback to the user. Regarding point 2), different versions of pattern detection algorithms are applied utilizing CPU, GPU and color processing if applicable.

The paper is organized as follows. Sect. 2 presents related works and paper contributions, in Sect. 3 an overview of adopted methods is presented. Experimental results are provided in Sect. 4. The paper is summarized is Sect. 5.

## 2    Related Work

Sensor calibration is a common first step in organization of many robotic or computer vision systems. The sensor model typically consists of intrinsic parameters as well as extrinsic parameters. Intrinsic parameters describe internal parameters of a sensor (e.g. sensor size, focal length etc. for monocular camera), while extrinsic parameters describe sensor position and orientation with respect to some external coordinate frame.

This paper concentrates on calibration of two range sensors (such as Lidar, ToF camera, RGB-D sensor or camera stereo-pair). In such case, we can safely assume, that intrinsic parameters of both sensors are either already calibrated or provided by the manufacturer, so we are able to obtain a valid 3D point cloud from each sensor. Thus, in a common scenario, the primary goal is to extract a common reference frame for both sensors.

A major problem in the calibration of sensors of different types (Lidar, ToF camera, RGB-D camera, camera stereo-pair), and of very different characteristics, is the extraction of the same pattern features in both sensors. In [6] and [11] flat patterns of characteristic shape are used. In [12] ball patterns are utilized together with RanSaC. Another direction is the application of flat patterns with holes [2,16], which enforces depth discontinuities in Lidar scanlines. Note also that both [12] and [2] exploit spatial structure of Lidar scanlines. The checkerboard patterns [5] are used less frequently due to difficult handling of Lidar data, but may be an option when a monocular camera is involved in calibration. Another possible approach is to resign from calibration pattern at all but treat the whole scene as a data source for features (e.g. lines in [10]).

In this paper a calibration pattern (a ball) similar to the one from [12] is used. A contribution of this paper is a research focus on computational efficiency of algorithms which is required for better user experience and system feedback during calibration. This goal is achieved by application of speed-ups to the process of detection in the form of color-based image pre-filtering or GPU-supported RanSaC. To achieve this, an existing PCL [15] implementation is extended and adapted to the problem discussed here.

## 3    Methods

The outline of the proposed calibration solution is given in Fig. 3. Two distinct paths are taken depending if the point cloud provided is an RGB-D cloud (an organized cloud with RGB field) or if the cloud is a generic XYZ cloud (unorganized cloud with no color information). In the former case the color segmentation is used first before applying RanSaC algorithm to detect calibration pattern, whereas in the latter case distance and plane filtering is applied first to increase processing speed and robustness. After collecting enough number of centroids, centroid cloud alignment is performed to retrieve desired transformation between sensors.

Sections 3.1, 3.2, 3.3, 3.4, 3.5 and 3.6 cover basic building blocks for cloud detection including basic CPU RanSaC version and its GPU vectorized

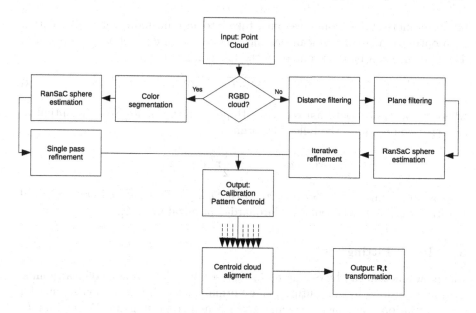

**Fig. 3.** Calibration procedure outline

counterpart. Section 3.7 covers calibration pattern detection in a generic XYZ cloud, while Sect. 3.8 covers calibration pattern detection in a RGB-D cloud. In Sect. 3.9 and 3.10 the last steps of the calibration procedure, alignment of cloud centroids and cloud pairing, are described.

## 3.1  Sphere Fitting

**Initial Estimation of Sphere Center and Radius.** Sphere center can be estimated using a set of at least 4 points in a non-singular configuration. A common method of extracting sphere parameters is to solve a system of four linear equations in the form

$$x^2 + y^2 + z^2 + ax + by + cz + d = 0 \tag{1}$$

where parameters $a$, $b$, $c$, $d$ are related to sphere center coordinates $x_0$, $y_0$, $z_0$ and sphere radius $r$ by

$$a = -2x_0,\ b = -2y_0,\ c = -2z_0 \text{ and } d = x_0^2 + y_0^2 + z_0^2 - r^2 \tag{2}$$

Having given 4 points on a sphere in a non-singular configuration we can create a system of 4-linear equations in the form of (1) and solve them for parameters $a$, $b$, $c$, $d$.

**Fine-Tuning of Sphere Parameters.** The estimation from only 4 points will provide a noisy solution to the problem of sphere parameters estimation. In order to obtain more stable results we must take into account more points. A solution is to optimize a raw Euclidean distance between a point and a sphere surface. This distance can be represented by the following residual

$$r(x_i, y_i, z_i) = \sqrt{(x_i - x_0)^2 + (y_i - y_0)^2 + (z_i - z_0)^2} - r \tag{3}$$

where $x_i$, $y_i$, $z_i$ are coordinates of the $i$-th point in the cloud. Now the optimization criterion assumes the following form:

$$\Phi(x_0, y_0, z_0, r) = \frac{1}{2}\mathbf{r}^T\mathbf{r} \rightarrow \min \tag{4}$$

This is an example of a *Non-Linear Least Squares Problem* and can be solved iteratively using e.g. Levenberg-Marquardt optimization [8,9].

### 3.2  Plane Fitting

Plane parameters can be estimated using a set of at least 3 non-colinear points. This can be done by computing plane normal vector first and then solving for the translation component. Having given 3 non-colinear point $P_1$, $P_2$ and $P_3$ the following cross product can be computed to obtain plane normal vector $\mathbf{n} = [n_x, n_y, n_z]^T$

$$\mathbf{n} = \vec{P_1P_2} \times \vec{P_1P_3} \tag{5}$$

The canonical form of plane equation is then computed as

$$n_x x + n_y y + n_z z + d = 0 \tag{6}$$

where $d = -n_x X_1 - n_y Y_1 - n_z Z_1$, and $P_1 = [X_1, Y_1, Z_1]^T$.

### 3.3  RanSaC Algorithm

RanSaC [4] is a commonly utilized robust-fitting method. The main purpose of the algorithm is model fitting that is robust to outliers. In a computer-vision community the algorithm have been widely used as a general-purpose object detector for objects that can be described by a model with few parameters [3].

RanSaC is an iterative algorithm. In each iteration the input data $D$ is sampled for some minimum number of points necessary to estimate a temporary model of the object $M_{curr}$. The estimated object is then verified against the whole dataset and the subset of inliers $I_{curr}$ consistent with the current model $M_{curr}$ is estimated. If the number of inliers $|I_{curr}|$ is greater than the largest number of inliers obtained in previous iterations $|I|$, the selected model and selected inlier set are updated ($I := I_{curr}$ and $M := M_{curr}$). The process is iterated until the maximum number of iterations is reached.

The required number of iterations can be estimated based on a confidence score $p$ denoting the probability of 'hitting' the true object at least once in the whole search procedure. Let $N = |D|$ be the number of points in the set and $N_o$ the number of points in our object and let $w = \frac{N_o}{N}$ be the probability of

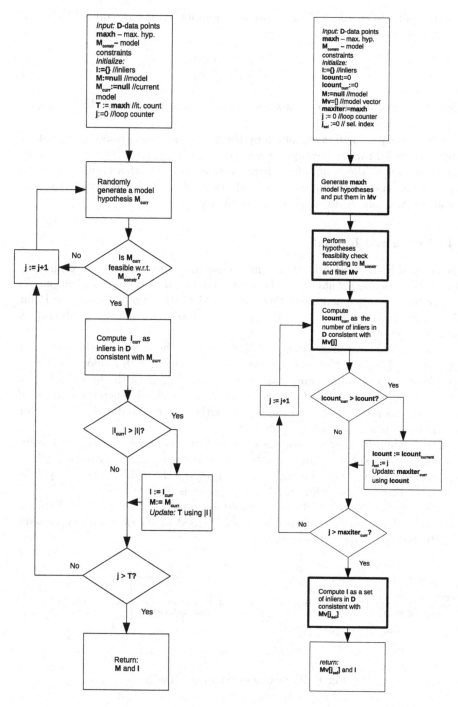

**Fig. 4.** RanSaC algorithm

**Fig. 5.** GPU RanSaC algorithm (vectorized actions are in bold)

selecting inlier in one draw. Then minimum number iterations required to obtain a confidence score $p$ is estimated by

$$T = \left\lceil \frac{\log(1-p)}{\log(1-w^n)} \right\rceil.$$ (7)

Since $N_o$ is typically unknown *a-priori*, so $w$ is estimated during algorithm execution using inequality $w \geq \frac{|I|}{N}$.

The model parameters estimated by RanSaC can be constrained to lie within some predefined range. This applies e.g. to the ball pattern with a fixed radius. Therefore, before the 'search-for-inliers' step, a feasibility check with respect to imposed model constraints $M_{constr}$ should be applied for better selectivity. The whole RanSaC procedure used in this work is given in Fig. 4.

### 3.4 Vectorized RanSaC

The original RanSaC algorithm is a time consuming operation with a complexity of $O(TN)$. The performance can be improved by GPU code vectorization. As a basis for the implementation there was used CUDA module from the Point Cloud Library [15] version 1.9.1 for easy integration with the Robot Operation System (ROS) [13].

The original procedure goes as follows: firstly, there is created a vector of hypotheses $Mv$ of length $maxh$. The vector is then filled with random candidate hypotheses using parallel computation. For each element of the hypothesis vector, the hypothesis is verified in parallel on the point cloud. Then the number of inliers is memorized. In the end the best hypothesis is restored and inliers vector is computed using a vectorized procedure.

The original PCL procedure was extended in this paper. Extensions involve introduction of a sphere model hypothesis together with feasibility testing, which is performed in a vectorized manner before applying the main RanSaC loop. The vectorized RanSaC procedure used in this paper is given in Fig. 5.

In the initial run of the RanSaC procedure $maxh = 10^5$ points is utilized for efficiency. If the procedure fails $maxh$ is extended to $maxh = 10^7$ hypotheses and RanSaC is re-applied (see: Fig. 6).

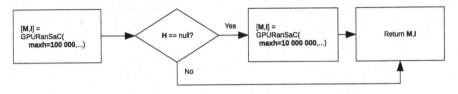

**Fig. 6.** Restarting of vectorized RanSaC

## 3.5    Single Pass Model Refinement

The basic RanSaC algorithm returns only some rough model estimate $M$ (computed using a minimum number of points) and a set of inliers $I$ in $D$ consistent with $M$. Since $M$ is obviously very noisy, some refinement is commonly used to fine tune model parameters. One option for the refinement is the estimation of the model using all available inliers. This we can call a *single-pass refinement*. Model refinement usually requires some non-linear optimization. An example of such refinement for a sphere model is given in Sect. 3.1 (par. Fine-tuning of sphere parameters).

## 3.6    Iterative Model Refinement

More computationally expensive but generally better procedure than given in Sect. 3.5 is an *iterative refinement*. The procedure repeats the following steps until the covergence of the inlier set $I$ is obtained:

1. Perform a *single-pass refinement* for the inlier set $I$ and update model $M$ accordingly,
2. Re-estimate the inlier set $I$ that is consistent with an update model $M$.

The procedure provides much better estimation of $M$. It also helps to filter out erroneous hypotheses (method usually diverges in such cases). Increase in computational complexity is to some extend mitigated by initial pruning of the cloud to the area covered by detected sphere (plus some safety margin).

## 3.7    Calibration Pattern Detection in XYZ Cloud

Extraction of calibration pattern center from a generic point cloud is divided into several stages:

1. Distance filtering. Close objects (such as robot elements) are filtered from the cloud.
2. Plane filtering. Iterative RanSaC-based procedure is carried out to remove planes from the scene (such as walls, floor or ceiling). The procedure stops when the reduction ratio (ratio between number of points in the detected plane and the original cloud size) is less than 0.1.
3. Estimation of calibration pattern center. The pattern center is estimated using RanSaC procedure for sphere detection.

The extraction can be performed using scalar processing (CPU) or vector processing (GPU) depending on the equipment available and user preferences.

## 3.8    Calibration Pattern Detection in RGB-D Cloud

In the case of RGB-D cloud, color is used to pre-filter areas of likely existence of the ball (calibration pattern). The input image $Im$ is first transformed to a

HLS (also known as HSL) color-space [7] to make solution resistant to changes in illumination. Then a thresholding is used to extract areas of promising color:

$$\mu_H - 3 \cdot \sigma_H \leq Im_H \leq \mu_H + 3 \cdot \sigma_H \tag{8}$$

$$\mu_S - 3 \cdot \sigma_S \leq Im_S \leq \mu_S + 3 \cdot \sigma_S \tag{9}$$

$$20 \leq Im_L \leq 220 \tag{10}$$

where all mu's and sigmas are established experimentally. Then, extracted regions larger than 100 pixels are evaluated starting from the largest to the smallest for the presence of a sphere. After preliminary RanSaC estimation of sphere parameters the hypothesis parameters are refined using method from Sect. 3.5.

### 3.9    Centroid Cloud Alignment

Let us denote the coordinates of the $i$-th ball centroid in the coordinate system of the 1-st sensor as $\mathbf{y}_i^1$ and in the coordinate system of the 2-nd sensor as $\mathbf{y}_i^2$. In order to find the best alignment between clouds we have to optimize the expression

$$\sum_{i=1}^{N} \mathbf{r_i}^2 = \sum_{i=1}^{N} ||\mathbf{R}\mathbf{y}_i^1 + \mathbf{t} - \mathbf{y_i^2}||^2 \tag{11}$$

with $\mathbf{R}$ being the rotation matrix $\mathbf{t}$ is the translation vector and $\mathbf{r_i}$ is the residual vector. This optimization criterion (11) can be solved in a single step using SVD method [1]. In the system proposed, the alignment is performed only when clouds meet certain geometric conditions. More precisely, two largest eigenvalues of the covariance matrix of centroids from the first sensor must be greater than 0.3 m.

### 3.10    Cloud Pairing

Pairing of clouds coming from different, not necessarily perfectly synchronized sensors, with different framerates is a considerable practical challenge. In the case of a production system there is used an Approximate Time synchronization algorithm [14] from ROS to solve this problem.

For the purpose of qualitative offline experiments a simple frame matching rule basing on frame timestamp is utilized. More precisely, two sets of frames are matched greedily one by one, starting from pairs having the most similar timestamp. The matching ends when all candidate matches are separated by at least 0.5 s (which turns out to be enough for a slowly moving operator).

## 4    Experiments

In the experimental section there is evaluated the accuracy of proposed algorithms (Sect. 4.1) as well as the computational efficiency of proposed solutions (Sect. 4.2). All experiments were performed using sensors mounted on a robot

owned, developed and provided by the United Robots company. The Lidar sensor
providing XYZ point cloud was Velodyne VLP-16, the RGB-D sensor providing
RGB-D cloud was IMU Intel Realsense D435i. The solution was implemented
with a support of PCL library v. 1.9.1 and integrated with the remaining part
of the robotic system using ROS framework [13]. Thrust library was used for
vectorized operations. The calibration system was tested both in a laboratory
environment (on a PC) as well as in production environment using NVidia Jetson
TX2 as well as NVidia Jeston Xavier. The main testing sequence consisted of
155 frames for each sensor.

### 4.1   Estimation of Calibration Error

Let us assume that measured centroid positions vectors $\mathbf{y}^1$ and $\mathbf{y}^2$ can be
regarded as samples from some probabilistic distribution which is modelled by
an isotropic Gaussian. Then for our experimental sequence the estimated residual standard deviation of a single centroid is $\sigma_0 = 0.044$ m (which is also equal to
the residual Mean Squared Error). The estimated errors for particular elements
of a parameter vector are given in the Table 1.

**Table 1.** Confidences of transformation parameters' estimation

| Param. | $\alpha_x$ | $\alpha_y$ | $\alpha_z$ | $t_x$ | $t_y$ | $t_z$ |
|---|---|---|---|---|---|---|
| | (deg.) | | | (cm.) | | |
| $\sigma_{param.}$ | 0.3855 | 0.4670 | 0.6913 | 0.56 | 1.22 | 2.15 |

Although the Mean Squared Error $\sigma_0$ is quite pronounced, uncertainties of
parameter estimations are kept in check due to utilization of adjustment method
for redundant data. Standard deviation for angle estimations does not exceed
$1°$. Uncertainties of translation are more pronounced but they are typically kept
at the level of 2 cm or less.

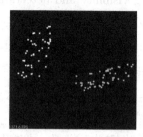

**Fig. 7.** Unaligned cloud of calibration
pattern centroids

**Fig. 8.** Aligned cloud of calibration
pattern centroids

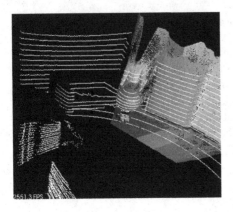

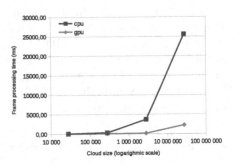

**Fig. 9.** XYZ and RGB-D cloud overlaid after calibration

**Fig. 10.** Performance comparison of CPU and GPU calibration pattern detection

One source of errors are inaccuracies of the RealSense sensor depth estimations. Another possible source could be imprecise time alignments between frames from both sensors. The unaligned clouds of centroids for our scenario are given in Fig. 7, aligned centroids are in Fig. 8. Figure 9 shows overlaid clouds from both sensors using estimated calibration parameters.

The error results reported here are moderate and comparable even to the results obtained using artificial data from [6] (in the cited paper the reported translation errors are 1–2 cm for translation and up to 0.6° for rotation component for quite moderate simulated noise levels).

### 4.2 Computational Performance

**Sphere Detection in XYZ Cloud.** Although the computational performance may not seem crucial for calibration (which can be performed off-line), a good near real-time computation is important at least at the stage of sensor data acquisition and calibration pattern detection. This enables to perform interactive calibration with immediate feedback regarding visibility and detectability of calibration pattern.

**Table 2.** Performance comparison of CPU and GPU calibration pattern detection

|  | GPU | | | | CPU | | | |
|---|---|---|---|---|---|---|---|---|
| Multiplier | 1 | 10 | 100 | 1000 | 1 | 10 | 100 | 1000 |
| Cloud size | 27 389 | 273 890 | 2 738 900 | 27 389 000 | 27 389 | 273 890 | 2 738 900 | 27 389 000 |
| Distance filt. (ms) | 0.19 | 0.59 | 1.66 | 9.12 | 0.33 | 4.90 | 61.10 | 692.19 |
| Plane filt. (ms) | 50.19 | 72.15 | 140.34 | 2123.11 | 23.72 | 249.60 | 2985.93 | 20832.60 |
| Sphere detect. (ms) | 15.35 | 13.70 | 27.80 | 124.43 | 57.94 | 74.81 | 658.56 | 4018.94 |
| Sphere refinement (ms) | 0.47 | 5.30 | 46.45 | 568.78 | 0.55 | 3.95 | 45.74 | 842.48 |
| Sum (w./o. ref.) (ms) | 65.72 | 86.43 | 169.80 | 2256.66 | 82.00 | 329.31 | 3705.60 | 25543.73 |
| GPU/CPU speedup | 1.2 | 3.8 | 21.8 | 11.3 | | | | |
| Sum (w. ref.) (ms) | 66.19 | 91.73 | 216.25 | 2825.44 | 82.54 | 333.27 | 3751.34 | 26386.21 |
| Refinement time share | 0.71% | 5.78% | 21.48% | 20.13% | 0.66% | 1.19% | 1.22% | 3.19% |

Therefore, in this paper there is provided a comparison of pattern detection performance using CPU and GPU processing for different cloud sizes. To conduct experiments, there was used a single frame of data containing 27389 points (from a Lidar Sensor). For subsequent tests the cloud was augmented by point multiplication (while maintaining the ratio between pattern and background points) to simulate larger data sizes that can be obtained from modern Lidar Sensors. There was compared a single-threaded CPU RanSaC implementation from PCL v. 1.9.1 and a vectorized GPU-RanSaC procedure discussed in this paper. The experiment was carried out on a computer equipped with AMD Ryzen 5 2600X Six-Core Processor and GeForce RTX 2070 GPU. Results of CPU/GPU comparison are given in Table 2 and Fig. 10.

The results provided demonstrate significant speed-up with respect to utilization of GPU in the calibration procedure for larger cloud sizes. The speed-up is smaller for the original cloud (of about 27 000 points), for larger clouds the speedup is greater, up to 21.8 for a cloud of about 2 700 000 points. Although speedup for the original cloud is small, it should be notable for clouds of around 500 000 points size (which is a value common for modern Lidars). For clouds consisting of up to several millions of points the processing time is near real-time and so the system could offer a good user feedback. This data size volume is realistic e.g. when accumulated clouds from multiple sensors are considered. The results given here are also important for some mobile devices (e.g. Nvidia Jetson and Nvidia Jeston Xavier), which offer much more GPU computation power than corresponding CPU power.

**Sphere Detection in RGB-D Cloud.** Detection of calibration pattern is several order of magnitude faster process for RGB-D cloud than XYZ cloud. The input cloud having size $640 \times 480$, which gives 307200 points can be processed in $5.03$ ms where color processing takes about $3.5$ ms, extraction of indices for object of interest takes $0.7$ ms and RanSaC $0.9$ ms on CPU. Due to color prefiltering, the RanSaC procedure returns a huge ratio of 98–99% of inliers and can be completed using 1–5 iterations.

## 5   Conclusions

In the paper there was demonstrated a system for calibration of Lidar and RGB-D sensors. The proposed calibration procedure uses a single ball calibration pattern moving over the scene and observed by a pair of calibrated sensors (Lidars or RGB-D sensors). Color filtering and GPU-RanSaC procedures provide near real-time performance on a desktop computer and ensure good interaction with an operator. The experiments show that this calibration setup offers accuracies comparable to those reported in literature for other calibration systems. In addition a significant speedup is observed due to utilization of GPU-supported procedures as well as proposed color prefiltering. Simulated experiments show that good computational performance is maintained even for large clouds, which makes the solution adaptable to the data density that will be offered by emerging new sensors.

# References

1. Arun, K.S., Huang, T.S., Blostein, S.D.: Least-squares fitting of two 3-D point sets. IEEE Trans. Pattern Anal. Mach. Intell. PAMI **9**(5), 698–700 (1987)
2. Debattisti, S., Mazzei, L., Panciroli, M.: Automated extrinsic laser and camera inter-calibration using triangular targets. In: 2013 IEEE Intelligent Vehicles Symposium (IV), pp. 696–701 (2013)
3. Fernandez-Rordiguez, S., Wedekind, J.: Recognition of 2D-objects using RANSAC. Technical report, Sheffield Hallam University, January 2008
4. Fischler, M.A., Bolles, R.C.: Random sample consensus: a paradigm for model fitting with applications to image analysis and automated cartography. Commun. ACM **24**(6), 381–395 (1981). https://doi.org/10.1145/358669.358692
5. Geiger, A., Moosmann, F., Car, Ö., Schuster, B.: Automatic camera and range sensor calibration using a single shot. In: 2012 IEEE International Conference on Robotics and Automation, pp. 3936–3943 (2012)
6. Guindel, C., Beltrán, J., Martín, D., García, F.: Automatic extrinsic calibration for lidar-stereo vehicle sensor setups. In: 2017 IEEE 20th International Conference on Intelligent Transportation Systems (ITSC), pp. 1–6 (2017)
7. Hasting, G., Rubin, A.: Colour spaces - a review of historic and modern colour models*. African Vis. Eye Health **71** (2012). https://doi.org/10.4102/aveh.v71i3.76
8. Levenberg, K.: A method for the solution of certain non-linear problems in least squares. Q. Appl. Math. **2**(2), 164–168 (1944). http://www.jstor.org/stable/43633451
9. Marquardt, D.W.: An algorithm for least-squares estimation of nonlinear parameters. J. Soc. Ind. Appl. Math. **11**(2), 431–441 (1963). https://doi.org/10.1137/0111030
10. Moghadam, P., Bosse, M., Zlot, R.: Line-based extrinsic calibration of range and image sensors. In: 2013 IEEE International Conference on Robotics and Automation, pp. 3685–3691 (2013)
11. Park, Y., Yun, S., Won, C., Cho, K., Um, K., Sim, S.: Calibration between color camera and 3D lidar instruments with a polygonal planar board. Sensors (Basel, Switzerland) **14**, 5333–5353 (2014). https://doi.org/10.3390/s140305333
12. Pereira, M., Silva, D., Santos, V., Dias, P.: Self calibration of multiple lidars and cameras on autonomous vehicles. Robot. Auton. Syst. **83** (2016). https://doi.org/10.1016/j.robot.2016.05.010
13. Quigley, M., et al.: ROS: an open-source robot operating system. In: ICRA Workshop on Open Source Software, vol. 3, January 2009
14. Ros approximate time synchronization filter. http://wiki.ros.org/message_filters/ApproximateTime. Accessed 14 May 2020
15. Rusu, R., Cousins, S.: 3D is here: point cloud library (PCL). In: 2011 IEEE International Conference on Robotics and Automation (ICRA), pp. 1–4, May 2011. https://doi.org/10.1109/ICRA.2011.5980567
16. Velas, M., Spanel, M., Materna, Z., Herout, A.: Calibration of RGB camera with velodyne lidar. In: Communication Papers Proceedings International Conference on Computer Graphics, Visualization and Computer Vision (WSCG) (2014)

# Video-Surveillance Tools for Monitoring Social Responsibility Under Covid-19 Restrictions

M. Sami Zitouni[1] and Andrzej Śluzek[1,2(✉)]

[1] Khalifa University, Abu Dhabi, UAE
mohammad.zitouni@ku.ac.ae
[2] Warsaw University of Life Sciences-SGGW, Warsaw, Poland
andrzej_sluzek@sggw.edu.pl

**Abstract.** The paper presents a number of recently developed methods for automatic categorization of socio-cognitive crowd behavior from video surveillance data. Sadly and unexpectedly, the recent pandemic outbreak of Covid-19 created a new important niche for such tasks, which otherwise have been rather associated with police monitoring public events or oppressive regimes tightly controlling selected communities. First, we argue that a recently proposed (see [31]) general socio-cognitive categorization of crowd behavior well corresponds to the needs of social distancing monitoring. It is explained how each of four proposed categories represents a different level of social (ir)responsibility in public spaces. Then, several techniques are presented which can be used to perform in real time such a categorization, based only the raw-data inputs (i.e. video-sequences from surveillance cameras). In particular, we discuss: (a) selected detection and tracking aspects for individual people and their groups, (b) practicality of data association combining results of detection and tracking, and (c) mid-level features proposed for neural-network-based classifiers of the behavior categories. Some illustrative results obtained in the developed feasibility studies are also included.

**Keywords:** Visual surveillance · Crowd behavior · Tracking · Automatic analysis · Covid-19

## 1 Introduction

In many countries (particularly in their cities and town) numerous surveillance cameras are installed in public spaces, in sensitive locations or in places with higher risk of dangerous behaviors. Even though such installations are often perceived as invigilation tools excessively used by police forces or oppressive regimes to control communities, they provide abundance of visual data on behavior of both individuals and human crowds.

Supported by Khalifa University.

L. J. Chmielewski et al. (Eds.): ICCVG 2020, LNCS 12334, pp. 227–239, 2020.
https://doi.org/10.1007/978-3-030-59006-2_20

Typically, those surveillance systems play a pivotal role (usually through a *post-factum* analysis) in monitoring public gathering, detecting anti-social behaviors of individuals and groups, providing warnings of security breaches, etc. Sadly, the recent pandemic outbreak of Covid-19 unexpectedly added a potentially novel dimension to these more traditional applications, i.e. monitoring social distancing (and other restrictions) in public places. Similarly to other tasks, social distancing monitoring has to be performed in dynamically evolving scenarios where individual humans and groups move, interact, merge, split, disperse, and execute other actions affecting the social structure of observed scenes. Obviously, monitoring should be preferably done in real time so that the violations can be detected and dealt with as soon as possible. With large numbers of surveillance cameras (e.g. in U.A.E. over 300,000 such cameras are installed in public places) automatic analysis of video-data is the only viable alternative to excessively large numbers of operators.

In this paper, we present several recently developed algorithms and methodologies which can be prospectively used in automatic surveillance of Covid-19 restrictions in public spaces. Although these tools have been generally designed for a wider task of socio-cognitive classification of crowd behaviors, we focus on their relevance to the social distancing monitoring problems. This is briefly discussed, and the corresponding conclusions are drawn, in Sect. 2. In the subsequent sections, we overview some results and recommendations in three main areas. First, detection and tracking for individual people and their groups in this context is presented in Sect. 3, including both traditional and CNN-based approaches. Then, in Sect. 4, practicality of data association by combining the results of detection and tracking is overviewed. Finally, Sect. 5 provides introduction to the socio-cognitive behavior classification based on mid-level features derived from detection and tracking results.

Final conclusions and observations are given in Sect. 6.

## 2    From Socio-cognitive Behaviors to Social Distancing

Automatic analysis of crowd behavior has been developed in the last 15 years, as reported in several surveys, e.g. [7,8,23,31]. Diversity of methods and algorithms indicate very complex relations between the raw-data images/videos and the classification methodologies. Thus, several proposals of systematic categorizations have been published, trying to address the problem from various perspective (e.g. visual information retrieval, [15,16], normal *versus* abnormal behavior, [2,11,25,26], crowds dynamics, [17,27], etc.).

In our recent study [31], we argue that socio-cognitive behaviors can be broadly categorize into four types:

1. *individualistic* behavior,
2. *social interaction* behavior,
3. *group* behavior and
4. *leader-following* behavior.

This categorization is not based on any psychological models, and may look unbalanced and asymmetric. For example, *leader-following* is actually a specific case of *group* behavior, while *social interactions* include a variety of diversified actions (from fighting to friendly gatherings). However, the survey of nearly 100 recent works suggests that such a typology well represents needs and tasks of visual surveillance systems.

Thus, it is not entirely surprising that social distancing can also be evaluated using this approach. We argue that the *individualistic* pattern of behavior is a more holistic and practical criterion than just keeping the nominal $2m$ distance, which is not always achievable in real-life situations (and difficult to estimate). In the individualistic behavior, any multi-person structures are accidental, quickly-disappearing, and people (even if instantaneously violating the nominal distance) do not stay in close proximity for too long.

In *social interaction* behavior, multi-person structures are formed more purposely and they last longer. This is a warning sign that social distancing might be violated, though not necessarily in a wide scale (e.g. only within a family or a group of friends). Finally, both *group* and *leader-following* behaviors usually indicate a serious violation of social distancing, and should be dealt with as quickly as practical.

(a)          (b)

(c)          (d)

**Fig. 1.** Examples of low-density (a) and high-density (b) group behavior, high-density individualistic behavior (c) and low-density social interaction behavior (d)

Examples in Fig. 1 further illustrate that simple distance measures (i.e. crowd density) might not properly reflect the actual behavior of the crowd, especially if extracted only from individual video-frames.

## 3   Detection and Tracking

Even though crowd analysis is used in a wide range of diversified problems, in almost all cases the low-level mechanisms of data processing are the same. In particular, detection and tracking of individual people and/or groups of people (and, rather infrequently, of body parts, e.g. heads) are the fundamentals of crowd surveillance, including socio-cognitive categorization. In this section, we highlight various aspects of these operations.

It is not our goal to provide a comprehensive survey. Instead, we focus on selected techniques (both conventional and based on deep learning) which are used in the performed feasibility studies.

### 3.1   Detection of Individuals and Groups

Detection of individual people in images/videos is a developed area (see [3]). It can be traced back to the pioneer work [4]. Subsequent developments, often

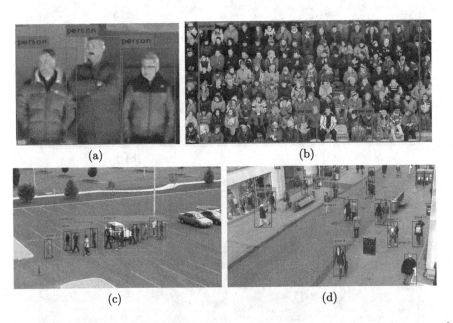

(a)                                           (b)

(c)                                           (d)

**Fig. 2.** Examples of good and imperfect performances by *state-of-the-art* detectors of people. Perfect (a) and unacceptably poor (b) results by YOLO detector [13]. Almost perfect (c) and only partially correct (d) results by ACF detector [5] (only blue bounding boxes represent detected individuals). Note that image quality is similar in all cases (Color figure online)

based on CNN, have been reported in numerous publications (e.g. [5,12,13,18, 30]). The results are almost universally represented by rectangular bounding boxes outlining the whole silhouettes or (especially in case of crowded scenes) only heads.

Nevertheless, the detection results are not always perfect so that they should not be considered 100% reliable. In Fig. 2, it is shown how selected *state-of-the-art* detectors may surprisingly fail because of various (sometimes inexplicable) reasons.

In this paper, we do not deal with detection of body parts, which is generally not needed in the discussed application (even though it may sometimes help to improve detection of individuals, e.g. [1,22,28]).

Detection of groups is a less advanced domain (e.g. [9,10,30]). The main reason is apparently a vague and subjective distinction between *group* and *non-group* configurations, which may affect both detection criteria and performance evaluation. The results are also typically represented by bounding boxes (e.g. green boxes in Figs. 2a, b), but binary patches (examples in Fig. 3) or intensity maps are also used.

**Fig. 3.** Group representation by binary patches

Figure 4 is a kind-of-summary example, where three CNN-based detectors (reported in [30]) are applied to the same video-frame to simultaneously detect

(a)    (b)    (c)

**Fig. 4.** Detection of individuals (a), small groups (b) and large groups (c) in the same frame, using three separate CNN-based detectors. Numbers represent confidence levels (outputs of the corresponding CNNs)

individuals, small groups and large groups. Simple intersection of the results provides an informative description of the crowd (including its hierarchical structure).

## 3.2  Tracking

Tracking (both individuals and groups) is necessary to understand the crowd behavior and evolution, especially from the socio-cognitive perspective. Obviously, the methodology of multi-target tracking has to be applied since multiple individuals and (often) groups are normally expected in the monitored scenes. For individuals, the problem consists in tracking well-defined (though usually non-rigid) objects; this problem has been investigated in numerous works (see the survey in [19]) and [14,20,21,24] are the exemplary papers discussing popular methods and algorithms of multi-target tracking.

Group tracking, similarly to group detection, is more difficult because of high volatility of tracked targets (including splitting, merging and disappearance) and smooth transitions between target and non-target objects (e.g. group forming or scattering). Specific techniques have been proposed (e.g. [6,10,29]) often based on the evolution of group outlines, tracking patches or even individual keypoints.

In our feasibility studies, we have tested various trackers, but eventually publicly available implementation of JPDA tracker [14] is the prime choice for tracking individuals, while groups are tracked using our own algorithm based on Kalman filter.

Regardless the quality of trackers, tracking multiple people (and, in particular, groups) in crowded scenes is seldom reliable because of partial/full occlusions, volatile visibility conditions, cluttered and non-stationary backgrounds, etc. Lost tracks happen frequently and re-initialization (i.e. incorporation of detectors into the tracking process) is often necessary. Therefore, the generated labels (identifiers of detected and subsequently tracked targets) not always accurately represent the actual identities of individuals and groups. Figure 5 shows an intricated pattern of group label changes during a very short period of tracking (just three subsequent frames).

(a)                        (b)                        (c)

**Fig. 5.** Examples of unexpectedly added new group labels (e.g. Groups 19 and 26) and group re-labeling (Group 14/15) over a very short period of time

In general, the processes of people/group detection and tracking in complex real-life scenes should be modeled stochastically. The corresponding probabilities represent, for example, detection certainty (in terms of actual existence and localization accuracy), certainty of data association in tracking (some targets may have alternative tracking options, e.g. in groups in uniformly dressed individuals), or the confidence level that a detected person is actually a member of a detected group (if individuals and groups are simultaneously detected).

The methodology improving uniqueness and continuity of labels assigned to individuals and groups is briefly discussed in the next section.

## 4    Data Association for Crowd Analysis

Labels (identifiers) assigned to individuals and groups play an important role in understanding crowd behavior and evolution. In particular, they can be used to clearly represent various processes, including: (a) patterns of group splitting and merging, (b) group size fluctuations, (c) switching group membership, (d) duration of group existence, etc.

However (as discussed in Sect. 3) reliability of labeling (and, thus, accurate representation of the relevant processes) is seldom satisfactory, mainly because of limited performances of detectors and trackers (even in good-visibility conditions). Therefore, we have proposed a probabilistic model of data association (i.e. a mechanism for creating, continuing and terminating labels) where the results of detection and tracking are complementarily used to maximize probabilities of correct label creation, continuity and discontinuity.

In general, we assume that the following probabilities can be estimated by detectors (i.e. at the current frame $t$) and by trackers (i.e. relating objects from the past frame $(t - K)$ to the current frame $t$).

A $p[i_n(t) \in g_m(t)]$, probabilities that $i_n(t)$ person is a members of $g_m(t)$ group (from detectors); they may indirectly include the probabilities of the actual existence of individuals and groups;

B $p[i_n(t) = i_m(t - K)]$, probabilities that $i_n(t)$ person is the same as $i_m(t - K)$ person (from trackers of individuals);

C $p[g_n(t) = g_m(t - K)]$, probabilities that $g_n(t)$ group is the same as $g_m(t - K)$ group (from trackers of groups);

In the trivial cases, these probabilities are just binary matrices of the sizes corresponding to the numbers of detected/tracked crowd components, but in practice (because of limited reliability of detectors and multi-target trackers) a continuous range of values is more typical.

By using these matrices (mainly max operations over their rows and columns) labels of individuals and groups can be efficiently managed. For example, a zero row/column in the corresponding matrix indicates a label termination or creation. Other typical situations (all of them important in automatic monitoring of crowd behaviors) which can be managed by this methodology include:

– Removal of non-existing crowd components (e.g. patches of moving background mistakenly recognized as groups or persons).
– Group merging and splitting (represented by linked labels).
– Evolution of group sizes and duration of their existence, etc.

Two diagrams in Fig. 6 illustrate how groups (represented by their labels) evolve in terms of their numbers, sizes and durations over two exemplary videos showing group behavior (Fig. 6a) and individualistic behaviours (Fig. 6b).

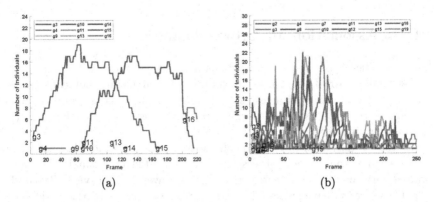

(a)    (b)

**Fig. 6.** Group behavior (a) with a limited number of large and long-lasting group labels. Individualistic behavior (b) with more group labels, but each group (even if instantaneously large) has a very short lifespan

To further improve these mechanisms, we have developed a data association formalism (based mainly of matrix multiplication) for estimating probabilities B and C from probabilities A (or another way around) so that detectors and trackers can be interchangeably used to substitute/supplement each other. Eventually, we obtained much more accurate and realistic estimates of crowd evolution patterns.

Statistics in Table 1 (obtained over a number of videos with individualistic and group behaviors) illustrate how the formalism improves the crowd analysis. We compare results based on the proposed association formalism against the baseline trackers/detectors. The comparison criteria include MAE (mean absolute error) in the number of group members, the average lifespan of group labels, the average numbers of labeled groups per video-frame, and the average total number of group labels created within a video.

MAE of the member numbers should always be low. Similarly, the average number of group labels per frame should be as small as possible to more compactly represents the crowd structure (Ockham's razor principle). However, a long lifespan of group labels (and, correspondingly, a smaller total number of labels) is a desired feature only in group behavior(where people actually move in groups). In case of individualistic behaviors, shorter lifespans of group labels (and, correspondingly, a larger number of labels) are more appropriate since groups continuously appear and disappear due to randomness of people movements.

**Table 1.** Comparison between group evolution results using only the baseline trackers and with the proposed data association formalism

| Behavior | MAE no of members | | Group label lifespan | | Group labels per frame | | Total no of labels | |
|---|---|---|---|---|---|---|---|---|
| | Form. | Base. | Form. | Base. | Form. | Base. | Form. | Base. |
| Group | 3.65 | 9.69 | 88.76 | 58.67 | 2.01 | 5.18 | 16.0 | 18.5 |
| Individualistic | 3.62 | 6.46 | 28.43 | 45.25 | 6.48 | 25.53 | 91.3 | 54.5 |

As shown in Table 1, all the above indicators significantly improve when the proposed formalism supplements results of the baseline algorithms.

## 5   Features for Behavior Classification

In the presented problem, the ultimate objective of the crowd analysis is detection of behaviors which are potentially irresponsible (i.e. group or leader-following behaviors; the actual distinction between these two classes is not really important), may indicate some violations (social interactions) or should be considered responsible (individualistic behavior). However, tools presented in Sects. 3 and 4 do not provide such results. They only convert the input video-data into (probabilistic) models and/or structural representations of processes taking place within the monitored crowds.

Our first classification methodology presented in [32] is based only on selected probabilities. The results on tested datasets are good. However, different behaviors are often identified in the same scene with very similar probabilities. This can be partially attributed to a general vagueness of crowd behavior interpretation (we can often perceive different behaviors in the same group of people) but also to a low utilization of data received from detection/tracking.

Behavior classification obviously requires a certain observation period. Unfortunately, data association results (even with improvements from Sect. 4) inherently deteriorate over longer sequences of video-frames. Therefore, our recent methodology is slightly different. We still propose input data from a significant numbers of video-frames (typically 20) which are used as inputs to a fully-connected shallow neural network. However, the actual feature values fed into NN are always obtained from at most two neighboring frames. Thus, if 20 frames are used, and each frame (and pair of neighboring frames) provides $T$ features, the total number of NN inputs is $T \times 20$.

The actual number of features extracted from a single (pair of) frame(s) is ten, i.e. $T = 10$, and they represent simple properties obtained from detection and tracking. Because tracking is performed only over two frames, the results are as credible as they can be. Some of the features are: the average speed of detected people (from trackers), the average group size and shape fluctuations (from detectors and trackers), the average number of group members (from detectors), etc.

We decided not to apply DL methods, which can often provide *end-to-end* solutions for vision-based classification tasks, because in the discussed problem such an approach is questionable. First, acquisition of sufficiently representative data is extremely difficult (if not impossible) for highly unpredictable appearances of people and groups in diversified/varying scenes, variety of physical behaviors for the same socio-cognitive behaviors, changing background and lighting conditions, etc. Even if a DL model can be trained, some of the learnt lower-level features would, for sure, somehow represent the detection/tracking outcomes which are already available (and easily adaptable to changing conditions). Our hand-crafted (after very tedious tests and trials) features have been elaborated over those detection/tracking results.

Compared to [32], the NN approach is much more discriminative. NN is apparently able to amplify factors (even if insignificant) differentiating between various behaviors, so that only the dominant behavior is given a high confidence value. Therefore, the method should be considered very robust to small changes of feature values (i.e., indirectly, to detection/tracking minor errors).

Illustrative examples are given in Table 2, where two video-sequences (one with individualistic and one with group behavior) are classified by both methods. Another spectacular example is given in Fig. 7, where (apparently because the alleged leader leaves the monitored area) the decision switches from *leader-following* to *group*.

**Table 2.** Results of NN-based approach *versus* [32] for the same sequences

| Behavior | Individualistic | Social interact. | Group | Leader-following |
|---|---|---|---|---|
| Seq.1 (NN) | **0.9907** | 0.0088 | 0.0004 | 0.0001 |
| Seq.1 ([32]) | **0.9224** | 0.7485 | 0.8002 | 0.0000 |
| Seq.2 (NN) | 0.0002 | 0.0054 | **0.8854** | 0.1089 |
| Seq.2 ([32]) | 0.8543 | 0.9315 | **0.9375** | 0.0000 |

Unfortunately, the *ground-truth* classification decisions are often subjective so that the exact performance evaluation is difficult. Nevertheless, we estimate that approx.90% of the socio-cognitive behavior classification results are correct in the performed experiments.

**Fig. 7.** The *leader-following* behavior is detected in (a) and (b), but in (c) the *group* behavior is identified (apparently because the leader left the monitored area)

## 6   Summary

The paper discusses a number of vision-based tools for crowd behavior analysis, focusing on socio-cognitive aspects. Even though the methods have been recently developed, it is just a sad coincidence that they become available when then need emerges for monitoring responsible social behaviors in the landscape shaped by Covid-19 restrictions.

By nature, the presented algorithms are not designed to stigmatize or penalize individuals. Instead, they just provide a general understanding of how responsibly people behave in the monitored public spaces. Further actions, if any, are beyond the scope of this paper. Nevertheless, certain extensions of the work (more focused on the individual crowd members) are anticipated. These would include, primarily, counting people wearing (or not) face masks, and detecting people with visual symptoms of infection (mainly coughing or sneezing).

## References

1. Alyammahi, S., Bhaskar, H., Ruta, D., Al-Mualla, M.: People detection and articulated pose estimation framework for crowded scenes. Knowl.-Based Syst. **131**, 83–104 (2017). https://doi.org/10.1016/j.knosys.2017.06.001
2. Baig, M.W., Barakova, E.I., Marcenaro, L., Regazzoni, C.S., Rauterberg, M.: Bio-inspired probabilistic model for crowd emotion detection. In: 2014 International Joint Conference on Neural Networks (IJCNN), pp. 3966–3973 (2014). https://doi.org/10.1109/IJCNN.2014.6889964
3. Chakravartula, R., Aparna, V., Chithira, S., Vidhya, B.: A comparative study of vision based human detection techniques in people counting applications. Procedia Comput. Sci. **58**, 461–469 (2015)
4. Dalal, N., Triggs, B.: Histograms of oriented gradients for human detection. In: 2005 IEEE International Conference on Computer Vision and Pattern Recognition (CVPR), pp. 886–893. IEEE (2005). https://doi.org/10.1109/CVPR.2005.177
5. Dollár, P., Appel, R., Belongie, S., Perona, P.: Fast feature pyramids for object detection. IEEE Trans. Pattern Anal. Mach. Intell. **36**(8), 1532–1545 (2014)
6. Edman, V., Andersson, M., Granström, K., Gustafsson, F.: Pedestrian group tracking using the GM-PHD filter. In: 21st European Signal Processing Conference (EUSIPCO 2013), pp. 1–5. IEEE (2013)

7. Grant, J.M., Flynn, P.J.: Crowd scene understanding from video: a survey. ACM Trans. Multimed. Comput. Commun. Appl. (TOMM) **13**(2), 19 (2017). https://doi.org/10.1145/3052930

8. Li, T., Chang, H., Wang, M., Ni, B., Hong, R., Yan, S.: Crowded scene analysis: a survey. IEEE Trans. Circuits Syst. Video Technol. **25**(3), 367–386 (2015). https://doi.org/10.1109/TCSVT.2014.2358029

9. Manfredi, M., Vezzani, R., Calderara, S., Cucchiara, R.: Detection of static groups and crowds gathered in open spaces by texture classification. Pattern Recogn. Lett. **44**, 39–48 (2014). https://doi.org/10.1016/j.patrec.2013.11.001

10. Mazzon, R., Poiesi, F., Cavallaro, A.: Detection and tracking of groups in crowd. In: 2013 10th IEEE International Conference on Advanced Video and Signal Based Surveillance (AVSS), pp. 202–207, August 2013. https://doi.org/10.1109/AVSS.2013.6636640

11. Mehran, R., Oyama, A., Shah, M.: Abnormal crowd behavior detection using social force model. In: IEEE Conference on Computer Vision and Pattern Recognition, CVPR 2009, pp. 935–942, June 2009

12. Raj, K.S., Poovendran, R.: Pedestrian detection and tracking through hierarchical clustering. In: International Conference on Information Communication and Embedded Systems (ICICES2014), pp. 1–4. IEEE (2014). https://doi.org/10.1109/ICICES.2014.7033991

13. Redmon, J., Farhadi, A.: Yolo9000: better, faster, stronger. In: 2017 IEEE International Conference on Computer Vision and Pattern Recognition (CVPR), pp. 7263–7271. IEEE (2017). https://doi.org/10.1109/CVPR.2017.690

14. Rezatofighi, S.H., Milan, A., Zhang, Z., Shi, Q., Dick, A., Reid, I.: Joint probabilistic data association revisited. In: 2015 IEEE International Conference on Computer Vision (ICCV), pp. 3047–3055. IEEE (2015). https://doi.org/10.1109/ICCV.2015.349

15. Shao, J., Change Loy, C., Wang, X.: Scene-independent group profiling in crowd. In: Proceedings of the IEEE Conference on Computer Vision and Pattern Recognition, pp. 2219–2226 (2014). https://doi.org/10.1109/CVPR.2014.285

16. Shao, J., Kang, K., Change Loy, C., Wang, X.: Deeply learned attributes for crowded scene understanding. In: Proceedings of the IEEE Conference on Computer Vision and Pattern Recognition, pp. 4657–4666 (2015)

17. Solmaz, B., Moore, B.E., Shah, M.: Identifying behaviors in crowd scenes using stability analysis for dynamical systems. IEEE Trans. Pattern Anal. Mach. Intell. **34**, 2064–2070 (2012). https://doi.org/10.1109/TPAMI.2012.123

18. Tomè, D., Monti, F., Baroffio, L., Bondi, L., Tagliasacchi, M., Tubaro, S.: Deep convolutional neural networks for pedestrian detection. Sig. Process. Image Commun. **47**, 482–489 (2016)

19. Wang, X., Li, T., Sun, S., Corchado, J.M.: A survey of recent advances in particle filters and remaining challenges for multitarget tracking. Sensors **17**(12) (2017). https://doi.org/10.3390/s17122707

20. Wen, L., Lei, Z., Lyu, S., Li, S.Z., Yang, M.H.: Exploiting hierarchical dense structures on hypergraphs for multi-object tracking. IEEE Trans. Pattern Anal. Mach. Intell. **38**(10), 1983–1996 (2016)

21. Yang, B., Nevatia, R.: Multi-target tracking by online learning a CRF model of appearance and motion patterns. Int. J. Comput. Vis. **107**(2), 203–217 (2014). https://doi.org/10.1007/s11263-013-0666-4

22. Yang, Y., Ramanan, D.: Articulated human detection with flexible mixtures of parts. IEEE Trans. Pattern Anal. Mach. Intell. (PAMI) **35**(12), 2878–2890 (2013)

23. Zhan, B., Monekosso, D.N., Remagnino, P., Velastin, S.A., Xu, L.Q.: Crowd analysis: a survey. Mach. Vis. Appl. **19**, 345–357 (2008)
24. Zhang, S., Wang, J., Wang, Z., Gong, Y., Liu, Y.: Multi-target tracking by learning local-to-global trajectory models. Pattern Recogn. **48**(2), 580–590 (2015)
25. Zhang, Y., Qin, L., Ji, R., Yao, H., Huang, Q.: Social attribute-aware force model: exploiting richness of interaction for abnormal crowd detection. IEEE Trans. Circuits Syst. Video Technol. **25**(7), 1231–1245 (2015). https://doi.org/10.1109/TCSVT.2014.2355711
26. Zhang, Y., Qin, L., Yao, H., Huang, Q.: Abnormal crowd behavior detection based on social attribute-aware force model. In: 2012 19th IEEE International Conference on Image Processing, pp. 2689–2692, September 2012
27. Zhou, B., Wang, X., Tang, X.: Understanding collective crowd behaviors: learning a mixture model of dynamic pedestrian-agents. In: IEEE Conference on Computer Vision and Pattern Recognition, pp. 2871–2878 (2012)
28. Zhou, T., Yang, J., Loza, A., Bhaskar, H., Al-Mualla, M.: A crowd modelling framework using fast head detection and shape-aware matching. J. Electron. Imaging **24** (2015)
29. Zhu, F., Wang, X., Yu, N.: Crowd tracking by group structure evolution. IEEE Trans. Circuits Syst. Video Technol. **28**(3), 772–786 (2018). https://doi.org/10.1109/TCSVT.2016.2615460
30. Zitouni, M.S., Sluzek, A., Bhaskar, H.: CNN-based analysis of crowd structure using automatically annotated training data. In: 2019 16th IEEE International Conference on Advanced Video and Signal Based Surveillance (AVSS), pp. 1–8, September 2019. https://doi.org/10.1109/AVSS.2019.8909846
31. Zitouni, M.S., Sluzek, A., Bhaskar, H.: Visual analysis of socio-cognitive crowd behaviors for surveillance: a survey and categorization of trends and methods. Eng. Appl. Artif. Intell. **82**, 294–312 (2019). https://doi.org/10.1016/j.engappai.2019.04.012
32. Zitouni, M.S., Sluzek, A., Bhaskar, H.: Towards understanding socio-cognitive behaviors of crowds from visual surveillance data. Multimed. Tools Appl. **79**, 1781–1799 (2020)

# Author Index

Printed in the United States
by Bookmasters

Printed in the United States
By Bookmasters